SMART Energy Management

A Computational Approach

SMART Energy Management
A Computational Approach

Krithi Ramamritham
Indian Institute of Technology Bombay, India
Sai University, Chennai, India

Gopinath Karmakar
Bhabha Atomic Research Centre Mumbai, India

Prashant Shenoy
University of Massachusetts, Amherst, USA

World Scientific

NEW JERSEY · LONDON · SINGAPORE · BEIJING · SHANGHAI · HONG KONG · TAIPEI · CHENNAI · TOKYO

Published by

World Scientific Publishing Co. Pte. Ltd.

5 Toh Tuck Link, Singapore 596224

USA office: 27 Warren Street, Suite 401-402, Hackensack, NJ 07601

UK office: 57 Shelton Street, Covent Garden, London WC2H 9HE

Library of Congress Control Number: 2021058266

British Library Cataloguing-in-Publication Data
A catalogue record for this book is available from the British Library.

SMART ENERGY MANAGEMENT
A Computational Approach

ISBN 978-981-125-228-0 (hardcover)
ISBN 978-981-125-229-7 (ebook for institutions)
ISBN 978-981-125-230-3 (ebook for individuals)

For any available supplementary material, please visit
https://www.worldscientific.com/worldscibooks/10.1142/12723#t=suppl

ॐ सह नाववतु ।
सह नौ भुनक्तु ।
सह वीर्यं करवावहै ।
तेजस्वि नावधीतमस्तु मा विद्विषावहै ।
ॐ शान्तिः शान्तिः शान्तिः ॥

(शान्तिपाठ, तैत्तिरीय उपनिषद)

Om Saha Naav[au]-Avatu |
Saha Nau Bhunaktu |
Saha Viiryam Karavaavahai |
Tejasvi Naav[au]-Adhiitam-Astu Maa Vidvissaavahai |
Om Shaantih Shaantih Shaantih ||
(Shantipath, Taittiriya Upanishad)

Om, May We be Equally Protected (the Teacher and the Student),
May We be Nourished together (with the fruits of learning),
May we perform (our studies) with Energy and Vigour,
May our Study be Enlightening, filled with Brilliance and not give rise to
Hostility,
Om, Peace (in me), Peace (in the environment) , Peace (in the forces
beyond physical and environmental).

Preface

Rapidly growing energy demand and our reliance on fossil sources have made energy a critical societal problem of the twenty first century. There is now an increasing emphasis to reduce our energy and carbon footprint by enhancing energy efficiency, reducing waste and adopting green renewable sources of energy. At the same time, technological advances in wireless networking, sensing and actuation, low-cost hardware and the Internet of Things (IoT) have made it possible to monitor and control all aspects of energy usage at unprecedented scales. The electric grid is being rapidly modernized in many countries through new technologies such as smart metering, sensing and actuation, automated grid management, distributed generation and energy storage. As a confluence of these trends, the use of novel ICT methods for the "greening of energy" has emerged as an important topic for researchers and practitioners.

This monograph provides a computational perspective to smart energy management, with an emphasis on smart buildings and the smart electric grids. The book emphasizes computational thinking and techniques such as inference and learning for smart energy management. To this end, this book is designed to help understand the recent research trends in energy management, focusing specifically on the efforts to increase energy efficiency of buildings, campuses, and cities.

The key topic emphasized in this book is smart energy management with the recurring theme being the use of computational and data-driven methods that use requirements/measurement/monitoring data to drive actuation/control, optimization, and resource management.

The book is intended to be a research monograph rather than a comprehensive textbook. It provides numerous case studies to elaborate on the covered topics. The target audience for this book includes researchers, graduate students, and practitioners. Researchers and graduate students will find this book useful for gaining an overview of recent advances in this area. Practitioners will find the book useful for understanding how technological advances can be put to practices and learn from case studies that bring out practical challenges in doing so. The book will also inform researchers from other domains, such as social and behavioral scientists who will gain an understanding of how technology can be used to change users' energy consumption behavior.

The book includes a discussion of many topics that either need additional research efforts before being assimilated into practice, or need a fresh look. This should help provide an impetus for further investigations in the important area of smart energy.

Krithi Ramamritham, Gopinath Karmakar and Prashant Shenoy
October 2021

Contents

Chapter 1

Introduction

Consumers expect electrical energy to be available whenever it is required, be it for charging a mobile phone or their electric vehicle, running kitchen appliances or office copiers, or for indoor climate control using individual air-conditioners or huge chilling units. But, rapidly growing energy demand and the dependency on fossil sources to meet the gross as well as peak demand have raised concerns over poor quality of service (occurrences of black-outs, brown-outs and load shedding), depletion of resources and impact on the environment. Even major developed economies, such as the USA, have experienced major power outages over the past decade [Amin (2007)]. The catastrophic blackout that India experienced in August 2012, which left more than 500 million people without electricity and basic amenities for several days, serves as a reminder of the urgency of acting on this challenge.

> Energy management is all about making energy available whenever it is demanded by the consumers in order to maintain certain quality of life and sustain growth to meet the human development goals. Energy management requires monitoring, controlling, and optimizing the performance of all the elements of the electric grid in order to provide the required energy of desired quality to the consumers.

Due to our desire to improve and maintain a certain quality of the environment in our homes and offices, not only is the demand for energy by buildings growing rapidly, but also the need for more robust systems for energy generation, transmission and supply to the end-user. Therefore, consumption, predictability of consumption, users' participation in the demand-response control are all going to play a big role in overall energy management. While the consumption in the industry can largely be reduced by improvement in the processes that results in efficient energy usage, current research shows that about 40% of the energy demand comes from buildings and there exists considerable scope for reduction in consumption and peak power demand. Both the overall consumption and the level of peak demand affect the electric grid and its performance — economic as well as electrical. For example, about 20% of the generating capacities exist in a power grid to meet the peak demand, which is used only 5% of the time [Farhangi (2010)]. In practice, quick-responding oil/gas fired based generating sources and hydroelectric plants are brought in to satisfy the peak demand. This is because they can be started within minutes and ramped up or down quickly to meet spikes in demand or sudden changes in the loads. While oil/gas turbine sources are inefficient and costly, the hydro generating sources have their own disadvantages — due to impact on the environment in terms of ill-effects on the land, impact on wild life, causing or aggravating flood situations. Therefore, the demand of the day is making both the grid and the buildings smarter by leveraging the recent developments in information and communication technology (ICT) making the sensors, actuators and the controllers smart.

The transitioning of existing electrical grids to "smart grids" involves a process that requires replacement of aging grid components, integration of renewable energy sources and energy buffering solutions, for example, storing excess energy in batteries, widespread deployment of sensors and actuators, and automating grid management using distributed Information and Communication Technology (ICT) systems [DOE 2012, Smart Grid Policy]. Concurrently, there has been an increasing focus on developing new technologies that will provide for a more sustainable future for our society. Since a significant portion of global energy use (and more specifically electricity use) continues to depend on traditional sources such as coal and natural gas, the use of novel ICT methods for the "greening of energy" has emerged as an important research area.

This is evident from two broad technology trends. First, there is a trend towards making the electric grid smarter, greener and more efficient. What makes a grid smart is the way this balance between demand and generation is maintained, mainly due to the focus on automatic detection of the imbalance between transmission and distribution and taking *preventive* actions rather than taking just *protective* actions against system failures. Whether a grid is smart or conventional, power generation, transmission and consumption must be kept in perfect balance in the system. Any imbalance will cause disruption in the quality of electricity supply in the form of blackouts, brownouts or load shedding.

We begin this chapter with an introduction to energy management we examine the fundamental characteristics of Energy management systems, introduce the basic terms, provide the necessary background and present the concepts related to energy. The need for a data and computation driven approach to energy management is motivated next. We end this chapter with a look at how the rest of this monograph is structured. In the process, we present our motivation behind writing this monograph and how readers can benefit the most from it.

1.1 SMART Systems

There is enormous excitement about synthesizing and benefiting from numerous technologies, including net metering, demand-response (D-R), distributed generation from intermittent sources such as solar and wind, active control of power flows, enhanced storage capabilities, and micro-grids. Additionally, since building energy use represents a significant fraction of total energy expenditures, a second trend is the design of smart residential and office buildings. These have the ability to interface with the smart grid and regulate their energy footprint, reduce peak consumption, incorporate local renewable energy sources and participate in demand-response techniques. The electric grid that results from these steps is said to be `smart`.

Consider a digital temperature sensor (a smart device — more than a simple sensor), senses the temperature of the surrounding atmosphere through its sensor (e.g., thermistor or thermocouple), its ADC (analog-to-digital-converter) circuitry samples (with the help of a processor — usually a microcontroller) the sensor's analog voltage output with some specified frequency and produces a digital value, computes/processes the digital data to find the equivalent temperature value and finally responds either with a display of the temperature on the LCD or sends data over the network. This device can also analyze the digital temperature data and compare it with a set value to generate an alarm if the temperature goes beyond the set limit. Here, we are essentially tracking the temperature and taking some simple actions.

For a more dramatic example, one with a lot of complexity, consider the systematic and fast evacuation of people during an emergency like fire, floods, etc., a significant concern in modern society. With the increased threat of attacks by miscreants, this has become even more important.

The problem has many dimensions. In case of fire and bombs in a building, the need is to quickly move people to the exits. In case of floods and water logging in a city/village due to disasters like Tsunami, people have to be moved to safe zones of the city/village. Since human life depends on the success of evacuation planning, a Smart Building Evacuation Planning System is required, which will help the building managers to evacuate people efficiently and systematically during an emergency such as scare, fires and terrorist attacks. In this case, sensors are deployed in the building to determine if a threat exists (such as a fire alarm). It will also use sensors to estimate the number of people present in the rooms and corridors of the building. We must also take into account the behavior of people during an emergency. Based on such information and the floor plans of the building, the system will suggest the routes that should be followed by building users during evacuations. Routes can be displayed using people's mobile phones, display boards and other notification mechanisms. The system is constantly executing the sense meaningfully–analyze–respond timely cycle until it is known that it has no more work to do, i.e., nobody is known to remain in the building.

In this example, we are sensing the environment carefully to determine whether or where humans may be present in the building, then depending on where people have been spotted we analyze possible solutions to choose the route for each person. We can extend such a building evacuation planning system to evacuate entire regions and cities. Of course, the scale of such an evacuation process will require highly efficient and scalable algorithms, along with low-cost and precise sensor technology.

Thus, a careful examination of the working of smart systems/devices/appliances reveals that they have a certain pattern in their behaviour: a device meaningfully senses the parameter that informs the system about the current state of interest, analyzes the sensed value (often after some processing) to help in decision making and finally produces a timely response, which can be a decision or a value. Let us look at the three phases that embody a smart system.

1.1.1 *Sense, Meaningfully*

Smartness of any control and monitoring system comes from accurate sensing of the environment and timely delivery of sensed data to the analytics subsystem for additional processing, analysis, control and further action. Sensor driven building management is motivated by goals like reducing and optimizing power consumption, monitoring the health of the building appliances, maintaining quality of the atmosphere in the building and tracking occupants in various parts of the building (useful for building safety and emergency evacuation), to name a few. Different sensing and inferencing mechanisms are used to obtain the observations pertaining to different facets of the building.

The accuracy of sensed data and latency of communicating it to applications determine the quality of service (QoS) of a system. The accuracy of sensing may be affected by faulty or biased sensors while timely delivery may be affected by queuing and processing of increased data traffic in the communication and computing infrastructure. Feeding inaccurate data for analytics or exceeding the latency bounds affects the performance of applications and thereby the reliability and responsiveness of the system. In practice, there is a tendency to over-provision sensors under the rationale that the more the data the more informed the decisions will be. Meaningful sensing relates to judicious sensing that ensures correct and timely decisions. Inaccuracy, unnecessary duplication or delays in sensing can make sensed values and hence decisions based on them meaningless.

A network of sensors is usually set up in the building by a BMS to obtain the information of interest. But installing these sensors in different parts of the building can a) be tedious and expensive b) cause inconvenience to the users c) increase the payback period and, d) affect the aesthetics of the building.

The fact that a sensor, suitable for observing a particular parameter, may in turn help to infer other parameters can be exploited to reduce the number of physical sensors deployed in a building. Similarly, inferences that are enabled by exploiting the structure of the building or the formal relationship between parameters, can lead to a better utilization of sensory resources.

1.1.2 *Analyze*

Analysis of the data sensed by the smart system has two major manifestations. One is based on archival or historical data. Another is on data pertaining to the prevailing situation. Given the online nature of decision making, i.e., we decide what is to be done in response to some real-world event, when the event happens, the response time is limited and hence we cannot always expect our choice of solutions to be optimal. Hence, often, to reduce the reaction or response time, the system analyzes the many possible solutions; the system state in which a particular solution will be appropriate will be analyzed and remembered by the system. When a real-time event occurs, the state that prevails then dictates the choice made.

1.1.3   *Respond, Timely*

Response is the action taken by the user or the system itself based on the analysis of the sensed data. Most "situations", unless reacted to in time, will escalate. Hence the response of the system should be timely, many of the situations will have timeliness related requirements (e.g., deadlines) attached to them.

Because of the above characteristics of the sense-analyze-respond cycle, in many scenarios, special purpose hardware is designed for one or more of the three phases.

Sense **M**eaningfully, **A**nalyze and **R**espond **T**imely

Aficionados of the English language will balk at the last part of this phrase. Still, we like to use this acronym given that it is very effective and *a propos*.

A SMART approach results when we have a smart control and monitoring system whose tasks will depend on the dynamics of the environment and whose responses are also situation dependent. Neither can be fully characterized statically (i.e., just a table lookup is insufficient to decide what to do at run time). Clearly, use of data from sensors to obtain situational awareness — state of both the environment as well as the system (resources) — is essential for meeting the challenges of such systems. Another crucial element is the synergy between the physical world and the ICT or cyber world.

What makes a grid smart is the way this balance between demand and generation is maintained, mainly due to the focus on automatic detection of the imbalance between transmission and distribution and taking *preventive* actions rather than taking just *protective* actions against system failures.

1.2 Computational Techniques in Energy Management

Computational Techniques in Smart Energy Management

The important concepts that have been borrowed from the domain of computer science in solving energy management problems and discussed in this monograph include the following.

- Real-time data communication and processing,
- scheduling,
- logic,
- algorithms,
- patterns,
- abstraction,
- optimization and
- machine learning.

It is interesting to note that a number of concepts from computer science inform us about how to solve some of the smart energy management problems. The example of an intrinsic computer science problem that finds its direct application in energy management is the communication and processing of huge amount of electrical grid data (voltage and current phasors and frequencies of the large number of interconnected buses geographically distributed over large distances) in *real-time*. This is essential for assessing the health of (the elements) in the grid and taking corrective actions in real-time (in the event of fault or excessive overload) so that blackouts and possible grid failure can be avoided.

The important concepts that can be borrowed from the domain of computer science in solving energy management problems include, i) Real-time data communication and processing, ii) scheduling, iii) logic, iv) algorithms, v) patterns, vi) abstraction, vii) optimization and viii) machine learning. Glimpses of applications of a few of these concepts follow.

As already mentioned, critical *real-time* grid monitoring applications require high data rate and strict latency. This demands designing of efficient query processing techniques that allow flexible bandwidth sharing among the applications.

Patterns can be observed in the energy consumption in homes/buildings due to appliances like air-conditioners (AC), washing machines and dishwashers. In case of on-off controlled ACs, the peak energy demand can be shaved by *scheduling* the operation of these appliances without compromising the thermal comfort requirement of the consumers. However, this requires *feasibility analysis* — derived from simple physics-based thermal models and *abstractions* of the building spaces with HVAC and *algorithms* for run-time control.

The consumption patterns of the appliances can also be used for non-intrusive load monitoring and finding the prevailing power usage. This information can be used for automatic intervention to achieve more economic use of these appliances by adjusting their time of operation. Inferencing *Logic* can be utilized as *soft sensors*, which can help in minimizing the number of physical or *hard* sensors. For example, from the temperature and RFID data of the occupants, the power consumption owing to ACs can be inferred. In case of deficiency in generation, power can be distributed *optimally* to the consumers using brownouts (rationing the available power to the loads based on their criticality/urgency of requirement for uninterrupted supply) instead of the existing practice of rolling blackouts. *Machine learning* techniques can be used for customizing the schedule-based HVAC control in commercial buildings based on dynamic adaptation of occupancy patterns.

In subsequent parts of this monograph, we illustrate the application of of the computational techniques used for solving problems in energy management.

1.3   Smart Electric Grid

In this section we will take a brief look at the inadequacies of the traditional grid and show how the modern smart grid is being designed to overcome these.

1.3.1   The Grid of the Last 100 Years

Historically, the electric grid has served as a common interconnection network connecting power generators with consumers. At any instance of time, all the generated electricity is consumed entirely. In other words, balance is always maintained between the amount of generation and consumption.

Generators are interconnected through a network of power transmission lines so that electricity reaches the consumer with the highest level of availability via these distribution lines. The grid can be viewed as a network of power transmission lines equivalent to links with generators as nodes.

The distribution lines facilitate tapping of electricity at various points on the *infinite bus*, namely, the grid, and make power available to consumers, both industrial and domestic.

The interesting property of the *infinite bus* is that ideally the electrical parameters, the voltage and frequency, of the grid remain unaffected even with changes in the electrical load connected to it. This is a necessity given that the loads are designed for particular voltage and frequency and their performance depends on the stability of these two electrical parameters. But, this does not mean that any amount of load can be connected or disconnected to the grid any time. The grid can accommodate fluctuations in load within its designed capacity. The stability of a grid under load fluctuations is briefly introduced in Chapter 2 and discussed with relevant details in Appendix B. Further, electrical loads consume two kinds of power: *real* and *reactive* (discussed in Appendix A) and therefore grid has to supply both types of power.

1.3.2 *Balancing Generation and Consumption*

Generation, transmission and consumption are to be kept in perfect balance in the electrical grid — smart or conventional. Any imbalance will cause disruption in the quality of electricity supply in the form of blackouts, brownouts or load shedding. The good old grid system is no longer adequate to meet the present requirements — mainly in i) offering support to diverse and large distributed generation, ii) monitoring grid health by handling large volume of data in real-time so that faults can be prevented, rather than mitigated and iii) facilitating consumers' participation in demand-response (D-R) control.

What makes a grid smart is the way this balance between demand and generation is maintained, mainly due to the focus on the automatic detection of transmission and distribution imbalance and taking *preventive* action rather than taking only *protective* action to system faults.

> In the *electric grid* balance between generation and consumption is maintained at all times. An imbalance can cause disruption in the quality of electricity supply resulting in blackouts, brownouts or load shedding. This balance between generation and consumption is a prerequisite for *normal* state of a grid.

1.3.3   *Peak Demand versus Aggregate Demand*

The demand for energy varies widely during the day. During certain hours in the morning and evening, demand is very high; such times are referred to as *peak* demand hours. Demand can become high during certain seasons like summer in tropical countries and winter in cold countries. Further, statistically, there can be sudden rise in demand in a grid with huge consumer base. In order to meet the peak demands, utilities must have provisions for additional generating capacities. Base load is managed by nuclear power plants and large thermal plants. But these cannot be brought in to meet sudden rise in demand, as it can take hours for them to start-up and get ready to be synchronized with the grid.

We have already discussed that quick-responding oil/gas fired generating sources and hydroelectric plants brought in to meet the peak demand are less economical and more hazardous for the environment. Thus most of the utilities penalize the consumers, especially bulk consumers by charging higher tariff during peak hours. Therefore, flattening of peak demand is a need for improving economic efficiency.

Smart Energy Management Tasks

- Providing power to consumers, ensuring quality — with greater availability at a lower cost.
- Enabling energy conservation — to decelerate the depletion of nonrenewable resources.
- Reducing the dependence on unsustainable energy sources — by avoiding unnecessary consumption.
- Increasing the use of sustainable energy sources — by exploiting renewables and finding ways to store excess energy from the sun or wind during periods of low consumption.
- Achieve peak shaving — by staggering loads or by scheduling appliances at the right times.
- Ensuring user convenience or comfort by automating tasks, providing timely feedback, or ensuring a comfortable environment.
- Incentivise users to become more energy efficient and adopt a more sustainable lifestyle.

Smart Energy management tasks require up-to-date information about the power required at a certain point in time, both for satisfying what is needed at that time instant and also for informed planning for the future. Which energy source to use should be decided for each customer or group of customers so that available energy is used optimally at all times, reducing collateral damage to the environment and providing energy of acceptable quality to all consumers. This implies that a "one size fits all" mindset or static decisions based solutions to energy management will not suffice. The disadvantages of the traditional approach to grid operations and energy management are further exacerbated by the inclusion of renewables in the source mix and the blurring of the classical distinctions between energy producers and consumers.

All of these imply the need for decisions based on knowledge about the current state of all the elements of the electric grid: Consumers, producers, and the interactions between them. This knowledge comes from the data made available from sensors embedded throughout the grid.

A smart grid uses the data available, from Phasor Measurement Units (PMU), to detect transmission and distribution imbalance vis-à-vis generation, overload conditions and takes *preventive* actions.

It exploits sensing, embedded processing and digital communications to enable the electricity grid to be

- *observable* (able to be measured and visualised),
- *controllable* (allowing it to be manipulated and optimised),
- *automated* (able to adapt and self-heal) and
- *fully integrated* (fully interoperable with existing systems and with the capacity to incorporate a diverse set of energy sources).

From what we have said thus far, we need up-to-date data to make the grid smart.

For example, QoS requirements for various grid applications demand data dissemination with timeliness guarantees. In addition, we also need all the consumers to be smart, that is, use the energy (made available to them) in a smart manner, and should design the generation decisions so as to be smart.

1.3.4   *Conventional Grid versus Smart Grid*

From the above discussion, it is clear that the scope of smart grids is very wide and therefore a short yet complete definition of smart grids is not easy. This is evident from the following definitions from the Joint Report [Giordano and Bossart (2012)] of European Commission (EC), JRC and US-Department of Energy titled "Assessing Smart Grid Benefits and Impacts: EU and U.S. Initiatives, 2012".

According to EC [EC Task Force for Smart Grids, 2010a],

A Smart Grid is "an electricity network that can intelligently integrate the behaviour and actions of all users connected to it — generators, consumers and those that do both — in order to efficiently ensure sustainable, economic and secure electricity supply".

According to the U.S. Department of Energy:

A smart grid uses digital technology to improve reliability, security, and efficiency (both economic and energy) of the electrical system from large generation, through the delivery systems to electricity consumers and a growing number of distributed-generation and storage resources.

Table 1.1 summarizes the key differences between the conventional grid and a smart grid. We will return to a more elaborate treatment of the smart grid in Chapter 2.

Table 1.1   Comparison between conventional grid and smart grid.

Topic	Conventional Grid	Smart Grid
Approach to power system faults	Detection and mitigation with a focus on protection of equipment	Focus is on prevention of fault by detecting emerging fault situations rather than responding only to the manifested faults.
System Monitoring and Control	Monitoring of grid health is limited to small number of large power plants and no real-time information for adaptive protection.	WAMS (Wide Area Measurement System) enabled by ICT to convey real-time information for improved monitoring and almost instantaneous stability of supply and demand on the grid.
Integration of renewable generation	Not equipped to support Distributed Energy Resources (DER)	Supports diverse and distributed generations with a focus on renewable resources
Power Quality (PQ)	PQ mostly neglected with focus on minimizing outages	Ability to identify and resolve PQ issues like voltage fluctuations, interruptions, waveform distortions prior to manifestation.
Consumers participation	Uninformed Consumers have no role to play in the power system management	Two way communication and active involvement facilitating deeper Demand-Response penetration

1.4 Smart Buildings

About 40% of total energy demand comes from buildings, commercial and residential, of which the major contributors are appliances for maintaining thermal comfort in buildings. Next to air-conditioners and heaters, the high energy consuming appliances in residential buildings, are washing machines and dishwashers. These appliances are often major sources of wastage of energy due to the absence of state of the art building energy management systems (BEMSs) combined with human negligence in not switching them off when not required.

A smart building refers to the new age building which provides better comfort levels to users, while minimizing energy consumption, handling safety and security issues, providing for maintainability, etc. It has an embedded BEMS in it to track and *control the use of available energy*, environmental parameters (e.g., temperature, humidity), occupancy status and count, etc. It is also able to reduce and optimize power consumption, monitor the status and health of the appliances in the building, profile energy consumption of different areas and identify zones with anomalous power consumption.

A smart BEMS is able to exploit various kinds of information which give deeper insights about the building. The required information comes from numerous sensors installed in smart buildings. Minimization of the number of sensors and use of *soft sensors* by exploiting the facet-sensor relationship (discussed in Chapter 3) is an example of extracting deeper inferable information thereby using fewer sensor data.

A building's energy footprint or its energy bills can be reduced, if wastage is reduced or prevented. Automated intervention by a BEMS can make a building *smart* so that energy can be saved and energy bill can be reduced without affecting the desired quality of life like provide thermal comfort with minimal dependence on human intervention. The main reasons behind wastage of energy in buildings are

- HVAC equipment, lights and fans remaining ON during periods of non-occupancy.
- Wrong placement of temperature feedback sensors, especially in auditoriums and large meeting halls, which can lead to over-cooling, to the extent that people start wearing jackets while inside an auditorium.
- Ad-hoc pre-cooling before the starting of an event like meetings, seminars etc.

- Lack of systematic and automatic health monitoring of equipment until it goes completely out of order.

Therefore, by detecting occupancy and opportunistically disconnecting loads in unoccupied rooms can save energy, a smart BEMS ensures that consumption is reduced and peaks are flattened.

Many electric utilities are moving away from a flat pricing model to a variable or peak usage-based pricing. In peak usage-based pricing, a utility monitors electricity usage over specific periods, such as every hour or every half hour, and bills customers, in part, based on the energy consumed in the peak period. So, a decrease in total and/or peak usage results in a more than linear reduction in the monthly electricity bill.

In sum, reduction in peak energy demand or prevention of energy wastage call for intervention by BEMS.

1.4.1   *Thermal Comfort in Buildings*

Thermal comfort in buildings is essential in order to provide favorable habitability and working environment in homes, offices, classrooms, auditoriums, etc. This requires provisioning of heating, ventilation, and air-conditioning equipment. Buildings are equipped with different kinds of air conditioning equipment like window/split AC, variable refrigerant flow (VRF) AC and large chiller plant depending on the types of spaces and the cooling capacity requirements. All these equipment/plants are energy consuming devices and they constitute 40% additional share of consumption in a building.

However, providing adequate cooling/heating capacity neither guarantees energy efficiency nor provides satisfactory thermal comfort to all the occupants. Complaints of over-cooling and under-cooling are common in offices, auditoria, etc.

Energy is often wasted simply by not following the discipline of switching HVACs, when not used. Preventing wastage is saving of energy. Therefore, automated intervention in operating HVACs based on occupancy becomes a necessity. Wi-Fi based occupancy sensing and schedule driven HVAC control (discussed in Chapter 5), where occupancy is estimated by monitoring and automatically learning the occupancy pattern, are examples of automatic intervention techniques. Another example is real-time chiller sequencing based on varying cooling loads, where historical data is used for prediction of COP (co-efficient of performance) using machine-learning techniques (Chapter 5).

Developing improved models that capture not only physical, i.e., thermodynamic, but also physiological, psychological, cultural and contextual factors, which play significant roles in the thermal perceptions of the individuals, is a challenge. Further, this also demands the model to be adaptive.

Therefore, providing thermal comfort to most of the occupants in an energy-efficient way is still a major challenge. We discuss the challenges and some solutions in Chapters 4 and 5.

1.4.2 *Solar Energy in Buildings*

The potential of integration of renewables like solar power with roof-top solar panels and building integrated photovoltaics (BIPV) is enormous. However, it has its own challenges, especially in the urban set up. The roof spaces are restricted by water tanks, AC outdoor units (ODU), DTH dishes, etc. In addition to space restrictions, the availability of direct sunlight on PV panels throughout the day is affected by these installations. Further, partial shading caused by nearby buildings, poles, overhead tanks and trees poses technological challenges in exploiting the full potential solar generation in buildings. These aspects of solar energy in buildings are discussed in Chapter 6.

1.4.3 *Smart Techniques for Handling Power Deficit*

In most of the developing countries like India, one of the nagging problems is deficiency in power generation to meet the demand. The prevalent approach adopted by the power utilities to deal with this problem is scheduled or rolling blackouts. In this approach, power supply in sub-areas within the distribution zones are disconnected (blacked out) for non-overlapping time intervals. On the other hand, consumers make their own arrangements for power during the blackout intervals. Commercial consumers resort to in-house diesel generators, which include small portable *gensets* (about 1 KW) used by shopkeepers and stores. Residential consumers use battery-backed inverters for their essential loads like lights and fans and the inverters are charged during the time slots when power is available. This adds up to the original problem of energy deficit as it involves charging and discharging efficiencies of inverter batteries. A brownout technique of managing building loads based on their priority by following the grid (available supply) is discussed in Chapter 7.

Another important area of research related to energy management in building is non-intrusive monitoring of loads, known as NILM. The NILM is attractive to the consumers as this is non-invasive and does not require installation of additional sensors. NILM is primarily aimed at collecting and analyzing data related to energy consumption and their pattern (power usage on different times of the day by different types of appliances) and thereby educate and influence individual consumers concerning the energy conscious utilisation of the appliances. NILM is an open research area and in Chapter 7, we introduce the fundamental techniques behind it, with examples.

1.5   About this Monograph

In this monograph we will focus on Computer Science approaches for addressing the problem of smart energy management. We believe that the so-called "computational thinking" can lead to approaches that are applicable to the topic of energy management. Further, data driven approaches and use of Artificial Intelligence (AI) techniques such as inference and learning also lend themselves to smart energy management. To this end, this monograph is designed to help understand the recent research trends in energy management, focusing specifically on the efforts to increase energy efficiency of buildings, campuses, and cities.

1.5.1   *Why this Monograph?*

Efficient use of energy is an age-old goal. But its importance has become even more apparent with the increased emphasis on human development and the increased use and thirst for more energy that it engenders. This monograph's *raison d'être* is its focus on addressing the energy concerns through the use of information and communication technologies (ICT). This has two implications: (i) Harness today's processing and communication tools to improve the efficiency and responsiveness of existing energy management systems. (ii) Use the ability of modern sensing and IOT (Internet of Things) devices to inform us about the current state of the system and provide a timely and state-appropriate (rather than a broad, imprecise) response, backed up by analysis. This goal will drive us to make use of recent research trends in data driven methods for improving energy-efficiency of buildings, campuses, and cities.

1.5.2   *Topics Covered by this Monograph*

Smart energy management in buildings — by users, and within the electric grid by grid operators, are the key focus areas of this monograph. There is enormous excitement about synthesizing and benefiting from numerous technologies, including real-time monitoring, net metering, demand response, distributed generation from intermittent sources such as solar and wind, active control of power flows, enhanced storage capabilities, and micro-grids. Additionally, since building energy use for maintaining thermal comfort represents a significant fraction of total energy expenditures, a second trend is the design of smart building management systems (BMS), which can maintain thermal comfort with improved energy-efficiency. Incorporation of local renewable energy sources is one of the techniques used to reduce dependence on traditional energy sources like fossil fuels. These obtain their smartness from being able to regulate their energy footprint, reduce peak consumption, and participate in demand-response techniques by cleverly using renewable energy sources.

This monograph is designed to help chronicle the recent research trends in energy management by examining various measures being undertaken for increasing energy efficiency of buildings, campuses, and cities and their connection with the smart electric grid. It examines the key enablers of smart buildings that will interface with the future smart grid. A common theme in our treatment of this broad area is the data-driven nature of the enabling technologies — we seek to analyze requirements, use measurement/monitoring data to drive actuation/control, optimization, and resource management.

Chapter 1 has already provided the motivation for managing electrical energy using smart (computational and data driven) techniques. Specifically, we motivated the need for being SMART from the perspective of deriving the benefit of computational techniques by means of facilitating interaction between physical and computational components.

Chapter 2 focuses on energy management issues within the smart electric grid. To this end, we discuss Phasor Measurement Units (PMUs), the sensors used within the grid, and show how queries over PMU data — required to track, manage and control the grid elements — can be deftly processed using the semantics of the data and the queries. We use examples from the Indian grid to drive home the advantages of this approach to data handling.

Chapter 3 is devoted to the study of buildings, viewed as smart systems. With this goal, we look at the SMART model in the context of buildings and identify the areas that demand attention to make a building smart. These are i) smart sensing, ii) modeling electrical loads, analyzing their pattern of operation and power consumption and iii) offering suitable control action to achieve the objective of reduction in energy consumption, flattening of peak demand and providing thermal comfort to the consumers. The section on smart sensing describes the gamut of sensors that have been and are being developed to obtain the building state with optimal deployment of sensor resources related to occupancy, power consumption, thermal conditioning, status' of appliances, etc., necessary to meet the objectives of smart buildings. Classification, modeling and analysis of electrical loads in buildings are discussed with a view to facilitating higher level energy optimizations such as flattening of peak demand and reduction in consumption. Real-world examples from buildings are provided throughout to exemplify various concepts.

Chapter 4 focuses on achieving thermal comfort for users. Heating and cooling are the dominant contributors to the energy consumption of buildings. Reducing the energy consumed due to heating and cooling while ensuring thermal comfort for building occupants is therefore a key challenge. The chapter discusses the various considerations in providing thermal comfort, factors influencing thermal comfort, the stages involved in providing thermal comfort given the lifetime of a building, undesirable phenomena requiring pro-active and reactive interventions along with a description of the many possible interventions. We also show the benefits of these interventions by creating a formal physics-based model for heat transfer in buildings, a model that is data driven. Case studies on thermal conditioning of different types of spaces conclude this chapter.

Chapter 5 explores the possibilities where thermal comfort can be customized to i) cater to individual preferences and ii) to prevent wastage of energy by occupancy-based control of HVACs, which includes maximizing the efficiency of the chiller plants under varying cooling-load.

Chapter 6 dwells on the potential of solar energy in buildings with a focus on roof-top solar PV and building integrated photovoltaics (BIPV) in urban areas. The associated challenges in mitigating the adverse effects of partial shading are also discussed along with technological solutions.

Chapter 7 discusses two more topics on energy management that involves both consumers and power utilities more directly. The first topic focuses on grid-following brownouts, which aims at offering a better and more acceptable solution than rolling blackouts during the energy deficient periods. The second topic relates to inferring various facets of a building's energy consumption in a non-intrusive manner, which aims at influencing and educating consumers towards energy-conscious use of appliances. It presents how data sensed by a single smart meter can be disaggregated into the constituent loads inside a building, a method known as NILM (Non-intrusive Load Monitoring). We describe various NILM methods that have been developed in the literature.

1.5.3 *What this Monograph is not about?*

This *book* is intended to be more of a "research monograph" rather than a textbook. The field of smart energy management has seen fairly active research in the last few years, but is still in a state of flux and many interesting problems remain. It is mature enough that some products have hit the market, but not stable enough to deter fresh startups from entering the arena. Given this, our hope is that this monograph will serve to spur further research and will help accelerate the cross-fertilization of ideas from multiple disciplines.

Given that the topic is interdisciplinary, our goal is to demystify ICT to people who study this problem from an electrical engineering or energy science and engineering perspective and for the IT and CS researchers and practitioners to be able to approach the energy issues with some comfort. But, clearly, we cannot delve deeply into the basics in these areas in a book of reasonable size, so much of the required background to the topics will be provided on a need basis. To bridge the gap, we will provide copious pointers to other literature and relevant background materials as appendices, which will help those interested in knowing more.

1.5.4 *Who Should Read this Monograph?*

The target audience for this monograph includes Students/ Researchers/ Practitioners interested in getting to know the latest developments in modern energy management using Computer Science tools and techniques driven by constraints imposed by the energy domain. These include the application of advance concepts from computer systems, analytics, hardware, networking, databases, etc., to address energy problems computationally.

Research students should find this monograph useful to come up to speed on what has been accomplished in the area and what problems remain.

Practitioners should find this monograph useful to check out solutions that are ready to be put into practice.

Researchers from other domains, e.g., policy makers and social scientists, who want to contribute to the spread and impact of the solutions being developed, and so want to understand the social angle, should also find the material accessible.

To this end, each chapter will start with a section that gives the necessary background and end with a summary which will bring to the fore the takeaways from that chapter, reiterating the learning and the gaps that exist in the state of the art/practice in that area. The bulk of the chapter will provide details of the developments in the topic covered by that chapter with numerous examples that will provide nuggets that carry the essence of the covered topics.

Chapter 2

Smart Electric Grid: Applications and Data Analysis

2.1 Introduction

The data communication network linking the transmission/ distribution grid monitoring/control infrastructure, advanced edge metering/monitoring/control infrastructure (AMI) in residential and commercial/industrial settings, and grid control and management points will be awash with data. A large number of monitoring and sensing devices, such as Phasor Measurement Units (PMUs), are being deployed throughout the network. Typical PMU data measured at a (generation or transmission) substation has three voltage phasors,[1] three current phasors, one frequency value, a GPS time-stamp, and other analog and digital values [IEEE Std C37.118.2].

In this chapter, various techniques for communicating data between phasor data concentrators (PDCs) are discussed. In the most elementary form, namely, centralized execution with unfiltered data forwarding technique (CEUT), all data is pushed towards the consumption points leading to large traffic which can adversely affect performance measures like *latency* and *bandwidth* of grid applications. It is observed that many applications running at super PDC, like bus angle monitoring, power system state estimation do not require this high data traffic. Thus, techniques are proposed in [Khandeparkar *et al.* (2017)] that perform application-specific in-network processing of data at local PDCs and send only filtered data to the super PDC. The mathematical expressions to estimate the processing latency at PDC in both centralized and distributed cases for any arbitrary hierarchical electric grid topology are also derived in [Khandeparkar *et al.* (2017)].

---

[1]A voltage/current phasor is a complex number that contains amplitude and phase angle information of the voltage/current.

Using these expressions, it is demonstrated that the total processing latency at PDCs of different levels in the Indian electric grid topology with proposed distributed approaches for each of the three applications (viz., bus angle monitoring (BAM), monitoring coherent group of generators (MCGG) and state estimation (SE)) outperforms the centralized approach.

Phasor Measurement Units (PMUs)

PMUs collect and transmit data that includes three voltage phasors, three current phasors, one frequency value and a GPS time-stamp. The emergence of the smart electric grid is driven by the increased need for achieving reliability, stability, support for distributed generation and integration of renewable sources.

In a smart grid, the signals are typically sampled and communicated at high rates — several 10s or 100s of times per second — to augment or even replace the conventional supervisory control and data acquisition (SCADA) in which measurements are done typically 4 times per second. A typical PMU data packet size is 100 bytes, but larger packet sizes are expected to be the norm soon as more and more grid parameters get included for monitoring. Thus, with PMUs having 90 phasors and 45 analog values at 100 Hz of sampling frequency, data from a single PMU could exceed 700 kbps.

High data rate and strict latency requirements of critical real-time monitoring applications in the smart grid require us to design efficient query processing techniques that allow for flexible bandwidth sharing among smart grid applications.

Various analysis tools and algorithms are used to aggregate and analyze the PMUs data from different geographical locations [23]. Phasor Data Concentrators (PDCs) at one or more levels aggregate and integrate the PMU data. Lower level PDCs (LPDCs) aggregate data from PMUs that are geographically located at different places (forming clusters), time align the data, and send the aggregated data to higher level or super PDCs (SPDCs).

Figure 2.1 shows a three layer hierarchical structure of planned electric grid [Navalkar (2012)] with GPS synchronized PMUs sending data to various PDCs.

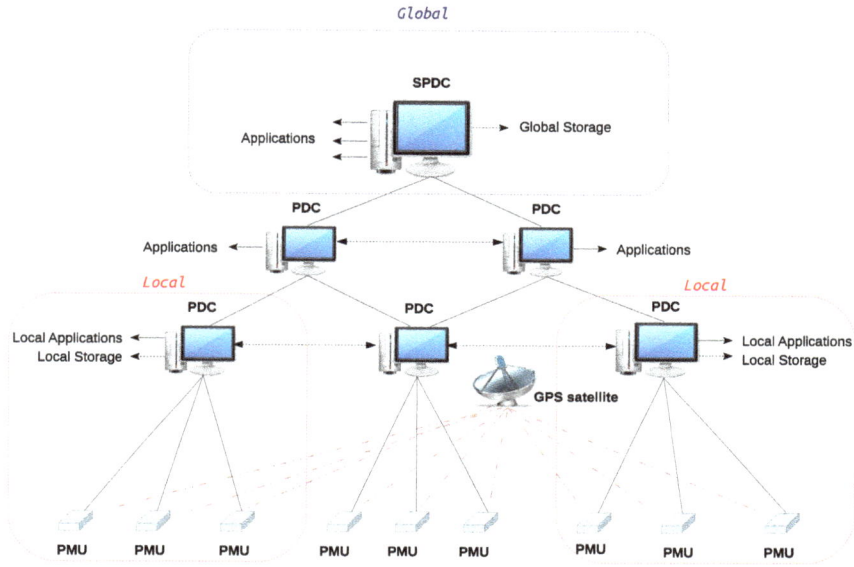

Fig. 2.1   The structure of the proposed Indian electric grid with PMU. [Khandeparkar *et al.* (2017)]

A general framework to devise efficient in-network query processing techniques for various applications in the smart grid is introduced in this chapter. In this approach, the application queries are converted into sub-queries to be executed at different PDCs. This framework allows for priority based application specific data dissemination in the smart grid. Using this, one can meet the QoS requirements of various applications while ensuring that reliability of the grid does not suffer.

The following scenarios arise in the context of executing a number of grid applications: Unlike the centralized approaches where all data are sent to a central PDC (also known as super PDC) in the grid hierarchy irrespective of applications that it runs, the distributed techniques send only application specific data. The query processing techniques systematically exploit application semantics to perform in-network processing at lower level PDCs and disseminate only filtered data packets to applications.

Specifically, we describe a distributed approach *DEFT* (***D****istributed **E**xecution with **F**iltered data forwarding **T**echnique*), which is semantics-aware — high degree of data filtering when the grid is stable, and the application receiving almost all the data when the grid is in danger of moving into instability. Novel approaches using estimation functions for PMU data are used to derive efficient in-network monitoring sub-queries based on them. The techniques also take cognizance priority of applications while concurrent applications are executing at upper-level nodes such that latency of time for critical applications does not suffer. Also, sending raw data from PMUs even at the lowest priority, enhances the reliability of the techniques to loss of packets in the network.

2.2   Sensing in the Grid, Meaningfully

2.2.1   *Phasor Measurement Units (PMUs)*

To harness data for making the necessary decisions in the smart grid a large number of monitoring and sensing devices, such as Phasor Measurement Units (PMUs), are being deployed throughout the network.

Typical PMU data gathered at transmission lines have three voltage phasors, three current phasors, one frequency value, a GPS time-stamp, and other analog and digital values [IEEE Std C37.118.2]. In a smart grid, the signals are typically sampled and communicated at high rates — several 10s or 100s of times per second — to augment or even replace the conventional supervisory control and data acquisition (SCADA), which acquires data from sensors every 2–4 seconds [Bobba *et al.* (2012)]. A typical PMU data packet size is 100 bytes, but larger packet sizes are expected to be the norm soon as more and more grid parameters get included for monitoring [IEEE Std C37.118.2]. Thus, with PMUs having 90 phasors and 45 analog values at 100 Hz of sampling frequency, data from a single PMU could exceed 700 kbps.

For India's grid, a total of 1600 PMUs is envisaged, making the total data size exceed 1 Gbps. These PMU measurements are synchronized using a GPS clock, enabling a consistent snapshot of the system. These provide the data necessary to assess grid performance and enhance the ability to control system operations and management.

Various analytical tools and algorithms are used to aggregate and analyze the PMU data from different geographical locations [Phadke (1993)]. PDCs at one or more levels aggregate and integrate the PMU data. Lower level PDCs (LPDCs) aggregate data from PMUs that are located at different places (smaller geographical locations forming clusters), time align the data, and send the aggregated data to a higher level or super PDCs (SPDCs). Figure 2.1 shows a three-layer hierarchical structure of the planned Indian electric grid [PGCIL (2012)] with GPS synchronized PMUs sending data to PDCs.

2.2.1.1   *Optimal Provisioning of PMUs*

As in the Internet, the grid power network offers many points of measurement and monitoring with different devices in use for monitoring the transmission and the distribution networks, generation sources, and building energy use. Significant investments are being made to deploy Phasor Measurement Units (PMUs) in transmission networks worldwide. The PMU placement problem — finding the minimum number and placement of PMUs to allow a bus system to be fully observable — is well-studied. The problem is often formalized as an Integer Linear Programming (ILP) problem [Xu (2004); Nuqui (2005); Dua (2008)], whose general solution is known to be NP-complete. In practice, several "real-world" considerations complicate the PMU placement problem. Because PMUs are expensive, it is typically not possible to deploy enough PMUs to observe all phasors [Ree (2010)]. In this case, it is desirable to place PMUs to observe a maximal number of buses, even if the network itself is not fully observable. One problem that has been recently investigated is the MAXOBSERVE problem — observing the maximum number of buses given a constant number of PMUs, showing that the problem is NP-complete [Gyllstrom (2011)]. Generalization of the MAXOBSERVE results has taken several directions, including the case when transmission links, as well as buses are observed, when there is a different value (utility) in observing different buses and links. PMUs are also subject to outages and failures, making the problem of cross-validating PMU measurements an important challenge. Cross-validation of PMU outputs can also help the development of other applications such as instrument transformer (CT, CVT) calibration. Algorithms for finding the minimum number of PMUs such that all PMU measurements are cross-validated and all system buses are observed have also appeared in the literature.

2.3   Analyze and Respond, Timely

Grid applications have varying quality of service (QoS) requirements, usually specified in terms of packets per second to be supported, data required, criticality of the application, tolerable data latency, geographic movement of data, and the deadline for bulk data transfer.

The breadth of data sources in the electric grid is matched only by the breadth of applications that will be consuming this data — the network operators, balancing authorities (BA), regional coordinators (RC), data archivers, state and federal monitoring agencies, and third-party applications such as demand response aggregators and energy markets may all desire access to portions of this data [Bose (2010)]. The communication network architecture and protocols that will support this vast array of data producers and consumers, with their varying priorities and performance requirements, will indeed be very different from today's centrally-polled, low-data-rate SCADA networks. Given the large number of smart-grid data producers/consumers and their heterogeneous requirements, a natural approach for organizing these elements into a coherent communication architecture is to adopt a publish-subscribe (pub-sub) approach for the smart grid [Gjermundrod (2009); Kim (2010)]. Such a pub-sub approach is preferable to application implementation directly over a TCP/IP (or SCADA) infrastructure since TCP's single-sender-to-single-receiver connection, byte-level reliability semantics, and fair congestion control are not well matched to application requirements [Hauser (2008); Kim (2010)]. More generally, a much richer set of communication abstractions and capabilities, far beyond traditional single-sender-to-single receiver communication, will be needed.

2.4   Smart Grid Applications: QoS Requirements and Background

The main goals of the smart grid applications are monitoring (the health of the grid), analysis and control during both normal and off-normal (in the event of fault) operating conditions. In addition, there are applications, which carry out post-incident analysis. A few important grid applications viz., bus monitoring application (BAM), monitoring coherent group of generators (MCGG) and power system state estimation (PSSE or simply SE), their QoS requirements and the necessary background information pertaining to electric grid are introduced in this section.

2.4.1   *QoS Requirements of Grid Applications*

As discussed in the previous section, the quality of service (QoS) requirements vary with grid applications. The importance of these grid monitoring applications and state estimation stems from the fundamental nature of the grid system, which always has to maintain a balance between generation of electricity by a large number of interconnected generators and the

Table 2.1   QoS requirements of grid applications.

Application	Latency (millisec)	Data items (Voltage: V, Current: I)	Data rate (packets/sec)	Time-Criticality
BAM	50	V angles	50	High
MCGG	50	V angles	50	Medium
SE	100	V, I magnitude and angles	25–50	Low

consumption of a huge number of electrical loads — industrial (plants with motors, pumps, valves, compressors, electrical furnaces including heating, ventilation, air-conditioning equipment (HVAC), etc.) as well as domestic (cooking ovens, refrigerators, washing machines, dishwashers, air-conditioners, etc.), connected at different geographical locations. *Energy is produced as and when it is demanded by the loads. It can only be stored as the rotational kinetic energy*[2] of the generators for a very short period of transition, when the load demand changes. As discussed, it requires monitoring of parameters and assessment of grid health in terms of its stability in maintaining the balance between power generation and consumption both during normal operation and in the event of fault in the power system. We will consider the following applications for discussions.

(1) BAM: *Bus Angle Monitoring*, and
(2) SE: *State estimation* of buses (also referred to as power system state estimation — PSSE).

In addition, we will also briefly discuss (in Appendix E) how the distributed analysis using efficient in-network query processing technique can help meeting the real-time processing requirement of *Monitoring Coherent Group of Generators (MCGG)*.

BAM and MCGG are important to analyze the stability of the grid so that corrective action can be taken to prevent its failure, which may lead to blackout. Note that in addition to BAM, state estimation is necessary to determine the stability of the grid and its monitoring.

We will discuss only BAM in this chapter. Two other applications are discussed in Appendix E for the readers who may be interested in them.

The key QoS requirements of the three grid applications viz., BAM,

---

[2]Here we stick to the core issue of inherent ability of the generators to handle transient imbalance in generation and consumption. This is to avoid unnecessary complexity in the current discussion by bringing in batteries and capacitors, which can offer some cushioning in the energy imbalance. However, that is a separate topic, but mentioned here for the sake of completeness.

MCGG and SE [Bakken *et al.* (2011); Phadke and Thorp (2010)] are listed in Table 2.1. It can be observed from the table that the time criticality of BAM is high and its latency requirement is low. The latency requirement of MCGG is the same as for BAM, but its time-criticality is medium. In case of SE, the time-criticality level is low and also the latency requirement is less stringent compared to BAM and MCGG.

2.4.2 *Background Information about Electrical Power Network/Grid*

In order to appreciate these grid applications, some background information about electric grid is necessary and therefore, a few facts about the electric grid are presented here. In addition, some basic concepts of electrical energy and the stability of electrical power system are briefly presented in Appendices A and B to facilitate the readers who lack the necessary electrical engineering background.

2.4.2.1 *Maintaining Balance between Power Generation and Consumption*

Consider a single steam turbo-generator[3] $G1$ connected to two induction motors $M1$ and $M2$. Note that an induction motor is an inductive load i.e., it requires both *active power* P as well as *reactive power* Q for driving any mechanical loads (e.g., pump, conveyor belt, etc.).

> Case I: *The motor $M1$ is connected to generator and the system is running in a perfect state of energy balance. Now, what if the load driven by a motor, i.e., the active power demand, is increased?*
> The motor tries to draw more power from the generator and the generator speed comes down momentarily. Sensing the fall in speed, the steam governor acts immediately to supply more steam so that the rated generator speed is restored commensurate with the increase in the active power demand. This is similar to a situation where a man is dragging a load at a constant speed. If some additional load is topped up, the speed of the man will reduce and he will have to put extra force to maintain the speed with the increased load.
> Case II: *The motor $M1$ is connected to generator and the system is running in a perfect state of energy balance. Now, what if the motor*

---

[3]The generator, which uses steam-turbine as its prime mover.

$M2$ *is also started i.e., both the active and reactive power demands are increased?*

The increase in the active power demand is taken care of by the steam governor valve. But, where from does the excess demand of reactive power come? The increase in the reactive power demand reduces the voltage output of the generator momentarily. This happens due to the demagnetising effect of the lagging current drawn from the generator, which reduces the magnetising field strength resulting in the reduction in voltage. Sensing the voltage, the automatic voltage regulator (AVR) increases the DC magnetic field excitation in the generator to supply more reactive power commensurate with the excess demand, and the rated voltage level is then regained.

Case III: *The motor $M1$ is connected to generator and the system is running in a perfect state of energy balance. Now, what if the load driven by the motor, i.e., the active power demand, is decreased?*

The motor draws less power from the generator and the generator speed increases momentarily. The steam governor valve acts immediately to reduce the steam supply so that the rated generator speed is restored commensurate with the decrease in the active power demand.

Case IV: *Both the motors $M1$ and $M2$ are connected to the generator and the system is running in a perfect state of energy balance. Now, what if the motor $M1$ is stopped, i.e., both the active and reactive power demands are decreased?*

The steam governor valve takes care of the decrease in the active power demand. The decrease in the reactive power demand causes an increase in the voltage output of the generator momentarily and the DC magnetic field excitation in the generator is reduced by the AVR to reduce its reactive power output in order to regain the rated voltage level.

Note that the above example is highly simplified and theoretical for the purpose of highlighting the basic principles on how a balance is maintained between generation and consumption, i.e., the load demand.

2.4.2.2 *Electric Power Grid: A Distributed Network*

Let us look at an abstract view of a power system presented in Figure 2.2. The figure shows that a power grid can be considered as a group of sub-networks, which are distributed geographically, away from each other and connected through *transmission lines*, called as *tie-lines*. It can also be

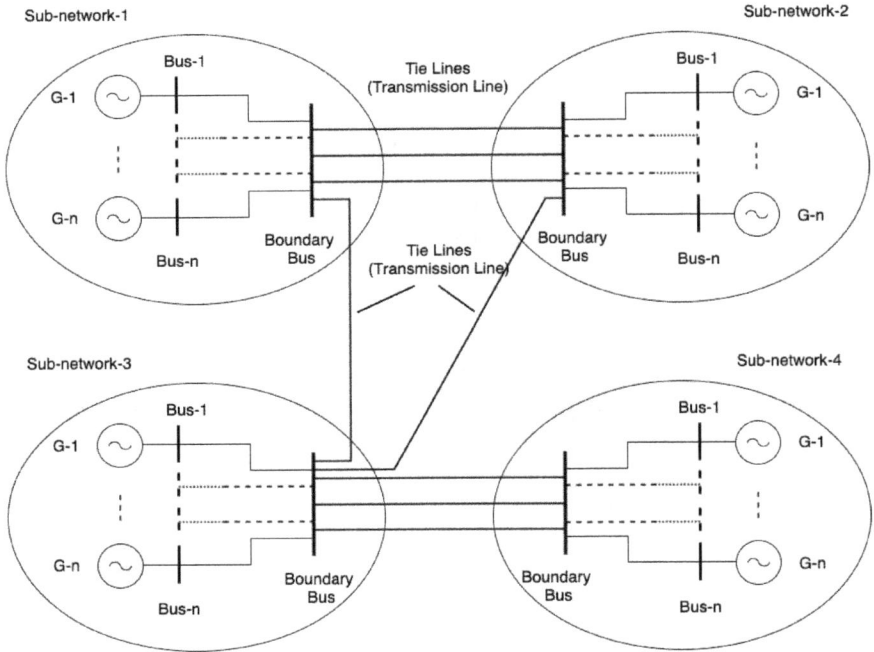

Fig. 2.2   Power Grid as distributed network of sub-networks.

seen from the figure that the interface buses that connect two or more sub-networks are called the *boundary buses*.

Today's large power systems pose practical limitations to model the entire power system in detail and this is usually not done in practice. Further, parts of the system geographically far away from the location of disturbance (electrical fault, tripping of generator or transmission line) have marginal effect on the system dynamics and therefore it is unnecessary to model those parts with great accuracy. For example, a fault in a sub-network in Figure 2.2 will have much larger impact on the generators within the sub-network. It may be noted here, in a practical grid, a large sub-network can consist of a number of small sub-sub-networks and each sub-network can have a hierarchy of PDCs.

2.4.2.3   *Power Flow in a Grid*

A larger number of generators are interconnected in a grid and so are the loads. A power grid can be divided into three groups or stages —

Generation, Transmission and *Distribution*. However, a practical grid is complex and the voltage levels of transmission and distribution vary widely and the various voltage levels of generation are stepped up to match the grid voltage using transformers.

We will concentrate on the power flow in a grid, which is important to understand the need for the grid applications under discussion. In an overhead transmission line, the flow of real power (MW) between two nodes depends on phase angle associated with the voltages of the nodes. This angle is termed as *load angle or power angle* δ and each node voltage is specified as $V\angle\delta$. In a system of interconnected generators, as in a power grid, this δ is always relative to a reference node voltage considered as $V\angle 0$. The power flows from larger δ to smaller δ. If two generator buses $G_1(V\angle 0)$ and $G_2(V\angle 30)$ are connected through a line, the power will flow from G_2 to G_1. When a distribution bus is fed by a generator $V\angle 30$ and its load increases, the rotor of the generator decelerates and gains its synchronous speed again when the governor feeds more steam to the generator to meet the increased power demand. However, its load angle increases in the process. For example, the new voltage phasor may become $V\angle 40$. Thus, *the difference in load angles between two nodes in a grid is a measure* of how much a particular transmission line or a distribution line is *loaded* (also called *stressed*) and therefore how far is it from the limit of handling flow of power.

2.4.2.4   *Grid Stability*

The knowledge of the load angles of the connected nodes in a grid is necessary to analyze the angular stability of a power system, which refers to the ability of the generators (synchronous machines) of an interconnected power system to get back into synchronism after being subjected to a disturbance. The disturbance can be the non-availability of a large generator or failure of transmission line/distribution bus due to electrical fault. In a nutshell, the difference in load angle (bus angle) across different buses in a power system is a measure of stresses across the grid and its propensity to instability [Hu and Venkatasubramanian (2007)].

Therefore, complete voltage angle and magnitude information for each bus in a power system for a specified load, the real power and voltage of the generators are essential for analyzing the health of a power system.

2.5   Data Dissemination and Grid Applications

In this section, we will explore semantics-aware distributed query processing approaches by carefully examining the applications mentioned in Section 2.4.

Semantics is of two kinds: *data semantics* and *application semantics*.

> *Data semantics* involves the physical meaning of data and its characteristics whereas *application semantics* involves domain knowledge about what the applications are trying to achieve.

For example, consider the bus angle monitoring application where phase angle difference is monitored so that its value is less than 2 degrees. For this application, it does not matter whether the actual angle difference is 0.5 or 1 degree as long as it is below the threshold. Application semantics also includes the data needs of the applications (voltage, current, and frequency), the data rate required by the applications, whether the application is a monitoring, control or protection application, etc.

2.5.1   *CEUT: Existing Approach of Data Dissemination*

The PMU data dissemination frameworks suggested in [Vanfretti and Chow (2011); Chenine and Nordström (2011); Armenia and Chow (2010); Adamiak *et al.* (2005)]:

> *send all the data from PMUs to higher level PDCs without considering their QoS requirements.*

In different embodiments, PMUs can send data directly to highest level SPDC where application queries are executing or PMUs can send data via lower level PDCs in a heirarchical manner.

Henceforth, we will refer to the PMUs sending all the data to SPDC via LPDCs as **C**entralized **E**xecution with **U**nfiltered data forwarding **T**echnique (**CEUT**). Figure 2.3 shows LPDCs forwarding all the data packets received from downstream PMUs to SPDC. In CEUT, the PDC waits for the arrival of data packets associated with the same time-stamp from PMUs till it times out. If a PDC does not receive all the data before the time-out, it sends whatever has been received to higher PDCs. An LPDC parses the input packet, extracting the measured values from the binary PMU data packet that are compliant to IEEE C37.118.2 format [IEEE Std C37.118.2]. For time-aligning the packets from multiple nodes, it temporarily buffers the extracted data, and once all the data with the same time-stamp arrive at

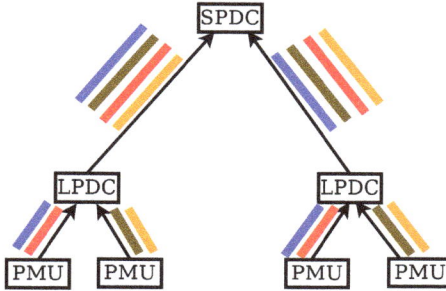

Fig. 2.3   CEUT — Centralized Execution with Unfiltered data forwarding Technique.

the LPDC, they are aggregated to create a single data packet. This data packet is then sent to the SPDC. The SPDC parses and time-aligns data packets from LPDCs and these are given to applications running at SPDC.

With PDCs at each level waiting for the arrival of PMU data packets to be aggregated and then sending to higher PDCs, a delay or drop of a PMU packet will cause the corresponding LPDC to wait until time-out. The overall delay from PMU to SPDC can ultimately *affect the QoS* of the smart grid applications. Moreover, pushing all the data packets from downstream PMUs/PDCs to higher levels PDCs, irrespective of the data required for applications, also *increases the bandwidth requirement.*

2.5.1.1   *The Need for Timeliness*

In order to showcase the importance of meeting timeliness requirements of certain applications, [Zhu *et al.* (2010)] simulate a real-world scenario with a fault of certain duration injected into a system. As the fault may result in oscillations that grow with time, making the system unstable, it has to be controlled within a certain duration in order to prevent cascading failure. The effect of different amount of delays in sending PMU data to a control application that is executed in a CEUT based fashion is studied in [Zhu *et al.* (2010)]. It shows that

> *beyond a certain time delay tolerated by the control application, the oscillations grow and the system becomes unstable.*

Hence, we need a *data dissemination framework with efficient algorithms* to disseminate the required data to the applications while ensuring their timeliness requirements.

2.5.2   DEFT: Data Dissemination with Timeliness Guarantees

We discussed CEUT approach which communicates raw data from PMUs to SPDC via LPDCs. As aforementioned, data dissemination with CEUT approach increases data traffic and hence, the processing latency, thereby affecting the performance of applications. Hence, the **D**istributed **E**xecution **with** **F**iltered data forwarding **T**echnique (DEFT) and its derivatives are presented as an alternative to CEUT. In this technique, the data delivered to the SPDC depends on the requirements of the applications running there. DEFT performs in-network processing of grid applications, thereby minimizing application specific data transfers and delays due to centralized computations. DEFT is an *application semantics*-aware processing and communication solution, in which the application is divided into coordinated sub-applications (sub-queries) executed in a distributed manner at LPDCs, and application relevant filtered and aggregated data gets forwarded from LPDCs to SPDC, as shown in Figure 2.4. When seen along with Figure 2.2, it can be observed that one or more PMUs send data to the LPDCs, which belong to individual sub-networks and each sub-network can have a hierarchy of local PDCs, depending on the complexity of the grid structure. Thus, this new distributed technique reduces processing overheads at LPDCs and the SPDC, by avoiding unnecessary data transfer between these nodes, which further helps in meeting latency requirements of analytics running at SPDC.

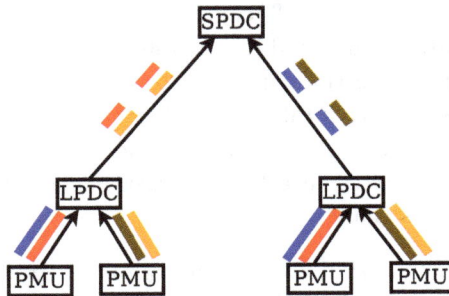

Fig. 2.4   DEFT — Distributed Execution with Filtered data forwarding Technique.

2.5.2.1   *Modeling PMU Data as Continuous Threshold Queries*

Grid applications over PMU data can be modelled as *continuous threshold queries* [Gupta and Ramamritham (2012)] over multiple data streams. A threshold, for example, could specify the maximum allowable phase angle difference. Proposed DEFT divides these queries into a number of sub-queries. For a phasor having frequency f, equal to that of the nominal frequency f_o (50 Hz for Indian system), the phase angle is constant. If the phasor has an off-nominal frequency ($f \neq f_o$) such that, f is fixed, the phase angle becomes a linear function of time. This relationship (discussed later in Section 2.6) is used to model the continuous sub-queries of BAM application that are disseminated at local nodes. The modelled subqueries are basically the local threshold conditions on the phase angles whose difference is monitored at a higher PDC. These conditions are derived from the query threshold and only when these conditions are violated, will then the data transfer to higher nodes occur. Threshold specified for the query is translated into thresholds for these sub-queries. Individual sub-queries involve data at a particular PDC, hence, can be executed at LPDCs. Results of individual sub-queries can be aggregated at SPDC to produce the result of the applications.

Continuous Threshold Query

- Threshold specified for a query is translated into thresholds for the sub-queries (e.g., in case of BAM, these are the local threshold conditions on the phase angles whose difference is monitored at a higher level PDC).
- Individual sub-queries involve data at a particular PDC and therefore can be executed at Local PDCs (LPDC).
- Results of individual sub-queries can be aggregated at SPDC to produce the result of the application involved.

Various methods exist for dividing a continuous aggregation query into a number of sub-queries. These sub-queries are executed at different nodes in the network. Sub-queries are derived in order to detect threshold violation(s). A query violation is asserted only if the threshold of one or more sub-queries is violated. In the event of query violation, it is important that an application in SPDC receives only the limited data pertaining to threshold violations to achieve the desired latency. On the other hand, it is also required that the SPDC, where the application query is executed, keeps on getting the data even when the threshold is not violated. However, it is necessary to ensure that no *false negative* case occurs. A false negative case is a scenario, where the application improperly indicates that no threshold violation has occurred, when in reality, it may have. Thus, it is non-trivial to have zero false negatives with minimum data transfer — so that the application gets all the data when needed and gets as little as possible when not needed.

In-network processing of smart-grid applications are categorized along two dimensions:

- Whether the sender node filters the data based on some condition, and
- Whether different nodes executing the sub-queries share information about each other or not.

BAM is an example where nodes executing a sub-query filter the data based on their local conditions. MCGG and SE applications are examples where the nodes executing sub-queries communicate, in this case, data from the common (boundary) buses. The boundary bus (refer to Figure 2.2) is discussed in Section 2.4.2.2.

2.5.2.2 *Prioritizing Data Dissemination and Processing*

A smart grid will have a number of applications running simultaneously, each with its own QoS requirements. As smart-grid applications have varying timeliness requirements, a priority based scheme is proposed in DEFT to assign more importance to time-critical applications compared to less time-critical monitoring applications. If we run these applications using DEFT, unlike CEUT where all data gets sent to SPDC, we can prioritize data for applications having more stringent timeliness requirements (e.g., BAM, which is essential for monitoring grid stability) compared to those having lenient QoS (e.g., SE). Therefore, a framework is proposed where a PDC processes data packets based on application priority by preempting processing of data needed by lower priority applications. Note that besides application specific filtered data packets, an SPDC may also require raw PMU data packets which, can be sent lazily. Multiple schemes are proposed to send raw data along with prioritized application specific data to SPDC.

2.5.2.3 *Combining Centralized and Decentralized Approaches*

Raw data from PMUs may be required at SPDC to analyze the impact of certain events such as load trips, generator trips, etc., on the entire system. All these can be made available at SPDC by combining Centralized and Decentralized approaches to improve generality and reliability.

The following two schemes are proposed that could be used in combination with priority based DEFT:

- DEFT+CEUT where, PMUs send raw packets to SPDC via LPDC. Here, DEFT offers prioritized and filtered data transfer and because CEUT is added with its data packets sent at the lowest priority, we get all the data at SPDC without interfering with the time-constrained application data.
- DEFT+CEUT-*direct* where PMUs send data directly to SPDC (i.e., bypassing LPDC), which processes them with low priority. Here, CEUT-*direct* provides an additional level of reliability by using different network paths for the raw data.

2.6 A Case Study on Data Dissemination for Bus Angle Monitoring (BAM) Application

BAM: Bus Angle Monitoring

Phase angle differences across different buses in a system are a measure of static stress across the grid and its propensity to instability. Thus, phase angle differences are required to be monitored with respect to predetermined stability thresholds.

It has already been discussed (Section 2.4.2.3) that power flow in transmission lines of an electrical network depends on the difference between the phase angles of the voltages at the two ends of the transmission lines. The power flow in a line, with voltages $V_t^s \angle \theta_t^s$ and $V_t^d \angle \theta_t^d$ at its two ends, is proportional to $V_t^s V_t^d \sin(\theta_t^s - \theta_t^d)$. Quiescent angular separation should be low to ensure safe operations of the grid. Following a disturbance (power system fault or tripping of a generator), oscillations in angular difference occur but are usually damped and equilibrium is restored. Angular instability occurs when the difference in phase angles between the ends of the transmission line increases uncontrollably [Hu and Venkatasubramanian (2007)]. Then synchronization between the two areas is lost — resulting in separation of the two areas, loss of generation units in those areas, and ultimately a blackout could occur. Thus, if the fault can be identified and rectified within critical *fault clearing time*, angular instability can be prevented. If θ_t^s and θ_t^d are the phase angles measured by PMUs located at buses s and d respectively and $\theta_{\mathbf{Th}}^{\mathbf{sd}}$ is the maximum phase angle difference allowed, an application correcting the consequences of fault does not need phase angle until:

$$\forall\, t \quad |\theta_t^s - \theta_t^d| \leq \theta_{\mathbf{Th}}^{\mathbf{sd}} \tag{2.1}$$

2.6.1 *CEUT: Centralized Execution with Unfiltered Data Forwarding Technique*

Bus angle monitoring with CEUT comprises of monitoring (2.1) at SPDC by extracting the phase angle information from the aggregated data packets sent by the LPDCs. Each LPDC in turn receives the data packet from the PMUs located at the two buses whose phase angles are being monitored at the SPDC.

The design and implementation of CEUT is motivated by [IEEE Std C37.244; Armenia and Chow (2010); Chenine and Nordström (2011)]. The basic functionalities of PDC are adopted from [IEEE Std C37.244]. In line with [Armenia and Chow (2010)], a dedicated thread is used to parse each PMU data packet at the LPDC. Further, the same thread performs the parsing and time-align operation. At LPDC, the time-align operation temporarily buffers the parsed data from multiple PMUs. Once the data packets from all the PMU with the same time-stamp are buffered, LPDC sends the aggregated data packet to the SPDC. At SPDC, the parsing and the time-align operations are similar to that of LPDC. There is a dedicated thread to parse data packet from each LPDC. The SPDC then delivers data to each of the applications.

2.6.2 *Performance of BAM with CEUT*

Performance of different queries in a variety of scenarios is evaluated using a combination of analysis as well as simulation. We first describe the data and then discuss different components of the application processing latency for LPDC and SPDC. We will also describe how the simulation results can be used to estimate processing latency for any application on the grid.

Table 2.2 Sample grid configuration.

Total buses	662
Total transmission lines	3006
Buses and transmission lines in Eastern region	363, 1532
Buses and transmission lines in Western region	299, 1469
Packet generation rate of each PMU	50
Angle difference pairs	14
Number of tie lines	5

2.6.2.1 *Data Generation*

First, simulation is carried out to generate data for the Eastern and Western regional grids of India. This data was then used to analyze the performance of applications in real-time in a separate simulation test-bed.

Table 2.2 shows the specifications of the simulated grid. The tie lines are the transmission lines connecting the two regions as represented in Figure 2.2. Together, the two regions have 662 buses with data generated at 50 packets per second for each bus (assuming PMU is located at each bus).

The primary performance metrics used are

- Average latency and
- Bandwidth usage

The latency is composed of the following.

- *Network latency* is the combination of latencies

    - from PMUs to LPDC and
    - from LPDC to SPDC

- *Processing latency* at LPDC and SPDC

As discussed, a smart grid containing 662 PMUs is simulated that cover the Eastern and Western regions of India. In this simulation, 14 pairs of buses were selected that directly or indirectly connect the two regional grids. The BAM application comprised of monitoring these 14 difference angle pairs at SPDC. The results show considerably lower latency, even in cases where some of the packets get dropped by the network. The simulated end-to-end latency from PMU to SPDC for BAM with CEUT was found to be 18.63 msec., of which 15.1 msec. was due to the combined processing latencies of LPDC and SPDC.

$$D_{ceut}^{lpdc} + D_{ceut}^{spdc} = 15.1 \ ms$$

where, D_{ceut}^{lpdc} and D_{ceut}^{spdc} denote the processing latencies of LPDC and SPDC, respectively.

Also, the bandwidth required from LPDC to SPDC with CEUT was 6.3 Mbps.

2.6.2.2   *Estimating Latencies at LPDCs and SPDCs for CEUT*

In this section, we give the expressions to estimate the processing latency at LPDCs and SPDCs for CEUT. The network latency is not estimated here. A network latency between the communicating nodes is mainly composed of the following.

> *Propagation delay*: It is the amount of time a packet takes to travel from a sender to a receiver and depends on the electrical behaviour of the communication medium and the distance between the communicating nodes.
>
> *Transmission delay*: It is the time required to shift all the bits of the data packet into the wire and is decided by the data rate of the link.

The backbone communication network planned for the Indian electric grid comprises of fiber optic cables [PGCIL (2012)]. However, as the length of cables between various communicating nodes and the data rate of each link is not specified in [PGCIL (2012)], we estimate the processing latency on the Indian electric grid from the observed processing latency on the testbed.

We conducted experiments using a simulation test-bed (mentioned in the previous Section 2.6.2) with a PDC processing varying number of PMU data packets of different sizes. The observed processing latency at the PDC was used to create a model, which can be used to estimate these latencies at PDCs of any arbitrary topology. The processing latency at a PDC consists of three components:

(1) D^{parse}: The time to parse incoming data packets,
(2) D^{agg}: The time to create an *aggregated* data packet using received data packets from PMUs/PDCs to send to higher level PDC, and
(3) D^{app}: The computation time of an application. Note that computation time is different for different applications e.g., BAM, MCGG and SE have different processing times.

For different PDCs (LPDC/SPDC) one or more of these components can be zero, as we will explain in this section. We start with giving expressions for the individual latency components first. Each PDC needs to process the incoming data packets from PMUs or PDCs at lower level in the hierarchy. As these packets are compliant to IEEE C37.118.2-2011 standard [IEEE Std C37.118.2], the first step in processing is to extract the phasor measurements and other field values from each data packet and populate this information in the objects of an internal data structure at PDC. This step is called parsing of the data packet. The data structure comprises separate variables to store the information present in the data packet such as *timestamp, phasors, analogs, digitals* etc.

For estimating D^{parse} and D^{agg}, we conducted two sets of experiments, Exp-1 and Exp-2. These experiments were done without any applications running on these nodes i.e., considering $D^{app} = 0$.

Exp-1 (Lowest Level PDC):

This experiment was done for lowest level PDC receiving data from a large number of PMUs (N). As N increases, we expect larger values of D^{parse} and D^{agg}. For a fixed data packet size, the value of D^{parse} is expected to grow linearly with the number of PMUs per core. Further, larger the data size, more would be the time spent by a thread to process the packet, and there will be more thread context switching overhead. Note that the data packet from each PMU is parsed by a separate thread at PDC; thread context switching occurs when a current thread has finished processing the data packet from one PMU, and a data packet from a different PMU arrives. In thread context switching, an operating system saves the registers and other private resources of the running thread, and loads the registers and private context of the next thread to be executed [Nichols *et al.* (1996), Ch. 6]. Hence, as LPDCs have more fan-in, there will be more threads. Further, larger the data packet size more will be the thread context switching time. *Thus, D^{parse} at an LPDC depends on the number of PMUs and the size of the data packet sent by each PMU.* The relationship between D^{perse} and data size K (KB) is presented in Table 2.3. Note that this is a limited set of data but sufficient (i) to establish the fact that the value of D^{perse} increases with data size, and (ii) to use it further for demonstrating the performance of bus angle monitoring using DEFT for various sizes of data. The value of D^{agg} can be obtained using Exp-2, as explained below.

Table 2.3 D^{parse} at LPDC for varying data sizes (K KB), per PMU per core in Exp-1.

K (KB)	0.1	0.5	1	1.5	2	2.5
D^{parse} (μ Sec)	124	289.6	360.48	407.2	477.6	564.4

Exp-2 (Intermediate PDC and SPDC):

This experiment was done for intermediate PDCs and SPDCs. These PDCs work with much lower fan-in. Hence, rather than the number of threads, the processing latency is dominated by the processing of lower level PDC data packet comprising of the maximum number of PMUs. In this case, D^{parse} was found to be 59.7 micro secs for each KB of data packet-size per PMU; whereas corresponding value of D^{agg} was found to be 43.6 micro secs for each KB of data packet-size per PMU. Thus, if an intermediate PDC receives largest data packet consisting of data from N_{max} PMUs with each PMU K KB of data in its frame, then its parsing and aggregation times will be as follows.

$$D^{parse}(K, N) = D^{parse}(t_{KB}) \times K \times N_{max} \qquad (2.2)$$

$$D^{agg}(K, N) = D^{agg}(t_{KB}) \times K \times N_{max} \qquad (2.3)$$

where, $D^{parse}(t_{KB})$ and $D^{agg}(t_{KB})$ denote the values of parsing time and aggregation time per t_{KB} of data, respectively. In this case, they are 59.7 and 43.6 respectively, as already stated. Similarly, D^{agg} for LPDC is also estimated to be $43.6KN$.

To summarize, the expressions for the estimated processing latency at LPDC and SPDC (or intermediate PDC) for CEUT are derived in this section. As explained earlier, one or more of D^{parse}, D^{agg}, and D^{app} can be zero for these cases. In the worst case, the data relevant for the application running at SPDC may be present in the largest data packet sent by the LPDC; and the data from all the LPDCs may reach the SPDC at the same time. The worst case processing latency at LPDC and SPDC in CEUT is:

$$D^{lpdc}_{ceut} = D^{parse}(K, N) + D^{agg}(K, N) \qquad (2.4)$$

$$D^{spdc}_{ceut} = D^{parse}(K, N_{max}) + D^{app} \qquad (2.5)$$

where N_{max} is the maximum number of PMUs such that their data are present in the lower level PDC data packets. There is no application specific processing at LPDC for CEUT, hence, D^{app} is not present in its expression. Similarly, an SPDC need not send aggregated data to any higher level PDC, hence, D^{agg} is 0 for it. Network latencies are ignored in these expressions.

Table 2.4   Observed and estimated processing latency at simulated LPDC (Western region PDC) and SPDC (National PDC) on a quad core processor ($C = 4$) and $K = 0.1$ KB.

	Observed Processing Latency (millisec)	Estimated Processing Latency (millisec)
LPDC	13.3	$(124*363/4)+(43.6*0.1*363)/1000 = 12.8$
SPDC	1.8	$59.7*0.1*363/1000 = 2.1$

2.6.2.3   *PMU Data Size for Simulated Eastern and Western Region Grids and Validation of Estimated Processing Latencies for PDCs*

The data sizes of most of the PMUs in the experiment varied from 80 bytes to 120 bytes. The mean and variance of PMU data size were 96 and 22 respectively. For simplicity, we consider the PMU data size as 0.1 KB (K), number of PMUs $N = 363$ and $N_{max} = 363$ (assuming one PMU per bus). It can be observed from Table 2.4 that the processing latencies — observed and estimated — for LPDC and SPDC are very close to each other.

2.6.3   *CEUT*-direct: *A Centralized Approach to Handle Loss of Data in the Network*

As discussed previously (Section 2.5.2.3), in addition to aggregated data from LPDCs, the applications running at SPDC may also be provided with an additional level of reliability. Accordingly, CEUT-*direct* is proposed, where PMUs multicast the data packets to both the LPDCs and SPDC. Thus, with CEUT and CEUT-*direct*, the SPDC would receive redundant copies of the same data items.

The average latency for BAM with CEUT+CEUT-*direct* was 20.4 millisec and was higher than CEUT (18.63 millisec). This is because SPDC had to process the aggregated data packets from the simulated LPDCs of Eastern and Western regions and the data packets from 662 PMUs of these regions. Replicated data also needs to be handled.

2.6.4   *BAM using DEFT (Distributed Execution with Filtered Data Forwarding Technique)*

As explained in the introduction of this Section 2.6, we need to monitor multiple buses and send an alert if the difference in the phase angles is above the specified threshold as given in Equation (2.1). For monitoring $\theta_t^s - \theta_t^d$, rather than sending all the phase values of θ_t^s and θ_t^d, we can assign thresholds to individual data values such that threshold θ_{Th}^{sd} is violated only if one or both of these thresholds are violated. However, unlike phase difference, the phase angles vary continuously with time. Hence, we need to model the phase angles and get the corresponding sub-queries with time-varying thresholds over individual phase angle value. The IEEE C37.118 standard [IEEE Std C37.118.2] prescribes that the PMU use multiples of the fundamental cycle (each cycle is of 20 millisec duration for a 50 Hz system) as the possible reporting intervals and the phase angle is reported between $\pm 180°$. The phase angle is constant when actual frequency f of the phasor is equal to the nominal frequency f_o. For off-nominal frequency ($f \neq f_o$) with a fixed frequency f of the phasor, the phase angle is a linear function of time and given by

$$\theta_t = [\Delta f \times t \times 360° + \theta_0] \, (mod \; 360°)$$

where, Δf is the frequency offset from nominal frequency in Hz, t is measurement time in seconds, and θ_0 is the phase angle at time $t = 0$. Using this data model, we obtain time varying thresholds over individual values of phase angles, which bound the phase angles with an upper bound and a lower bound.

Case-1: Frequency for both the buses are same.

If the frequency is the same for both the angles whose difference is monitored, these two bounds should be linear with time, parallel to each other, and the distance between them should be equal to the phase angle difference threshold (θ_{Th}^{sd}) as shown in Figure 2.5. We can derive the individual values of thresholds using the method described below. If the offset frequencies from the nominal frequencies for buses s and d, corresponding to angles θ^s and θ^d, are equal ($\Delta f^s = \Delta f = \Delta f^d$), the trajectory of individual values of the phase angles will be parallel to each other. Figure 2.6 illustrates this case with the initial assumption that θ^s is greater than θ^d and $\Delta f < 0$.

If β^s is the distance of the upper bound from linear equation estimating θ_t^s (line l_1) and β^d is the distance of the lower bound from linear equation estimating θ_t^d (line l_2), we get,

$$\beta^s + \beta^d = \theta_{\mathbf{Th}}^{\mathbf{sd}} - |(\theta_0^s - \theta_0^d)| \tag{2.6}$$

where $\beta^s \geq 0$ and $\beta^d \geq 0$ and θ_0^s, θ_0^d are the phase angles of buses s and d respectively at time $t{=}0$. From Figure 2.6, we know that the slope of the two parallel lines is Δf. The line equations for l_1 and l_2 are then given by,

$$l_1 : \theta_{est}^s = (\Delta f \times t \times 360) \, (mod \; 360°) + \beta^s$$
$$l_2 : \theta_{est}^d = (\Delta f \times t \times 360) \, (mod \; 360°) - \beta^d \tag{2.7}$$

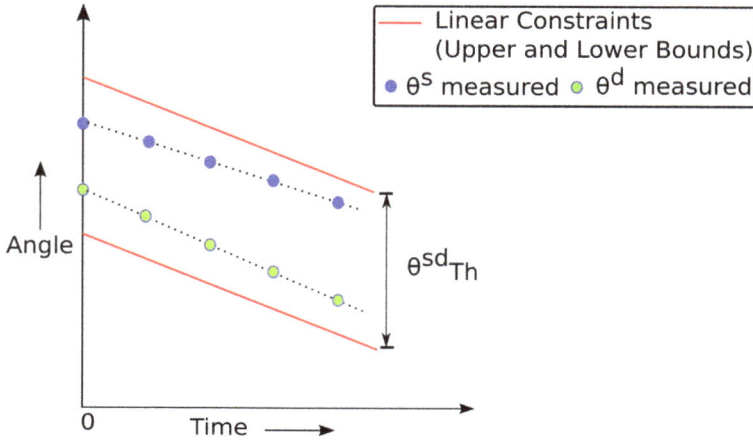

Fig. 2.5   Angle difference threshold monitoring technique.

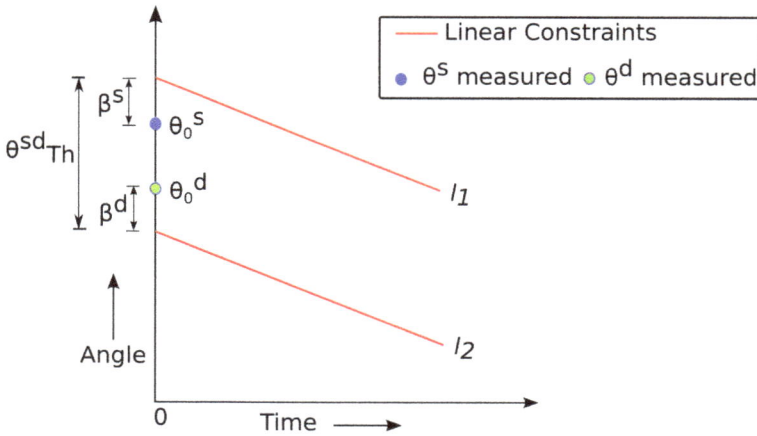

Fig. 2.6   Deriving initial constraints on the phase angles θ^s and θ^d.

where θ^s_{est} and θ^d_{est} are the estimated phase angles. Thus, constraints l_1 and l_2 can be set for both θ^s and θ^d at their respective LPDCs as follows,

$$\theta^d_{est} \leq \theta^d, \theta^s \leq \theta^s_{est}. \tag{2.8}$$

Rather than sending continuous phase angle information to SPDC (that monitors (2.1)), LPDCs can send data only when any one of the constraints

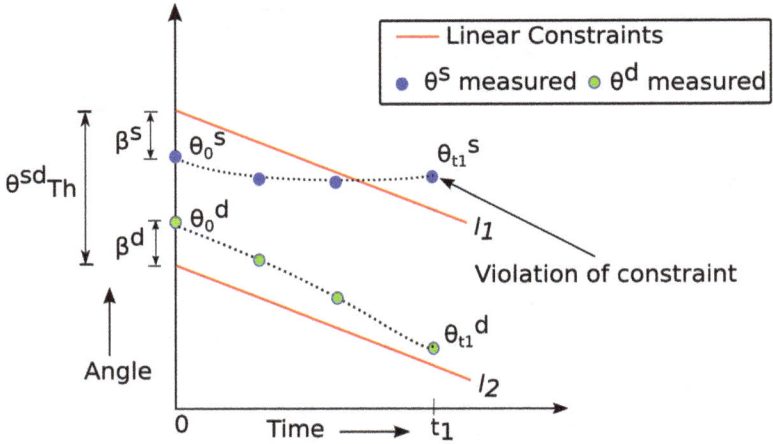

Fig. 2.7   Angle θ^s violates constraint l_1.

(2.8) *is violated.* Figure 2.7 shows that until time t_1, no data need to be transferred from LPDC to SPDC. At time t_1, θ^s violates constraint l_1 and θ^s value is sent to SPDC.

There are two ways to obtain the β values in (2.7). A naive approach is to equally distribute the offset, $\theta_{Th}^{sd} - |(\theta_0^s - \theta_0^d)|$ among θ^s and θ^d i.e.,

$$\beta^s = \beta^d = \frac{\theta_{Th}^{sd} - |(\theta_0^s - \theta_0^d)|}{2}.$$

A second approach obtains β values dynamically using the rate of change of frequency (ROCOF) information sent by PMU located at each bus [IEEE Std C37.118.2]. The larger ROCOF of a bus implies that the bus angle is changing at a faster rate and hence, its β value could be increased relative to that of the other bus while Equation (2.6) is satisfied.

Case-2: Frequencies for both the buses are different.

If the frequencies at the two buses are different, the upper bound and lower bound on the phase angles of the buses will not be the same. Hence we derive the equation of the linear bounding constraints based on the average frequency $(\Delta f_{avg} = (\Delta f^s + \Delta f^d)/2)$, i.e., use Δf_{avg} in (2.7) while dividing the β values in proportion to the corresponding rate of change of frequency values. Henceforth, we refer to violation of type (2.1) as a *global violation* and violation of (2.8) as a *local violation*.

Table 2.5 Data sizes of different messages for BAM in DEFT.

LPDC to SPDC		SPDC to LPDC	
Meta info	16 B	Meta info	16 B
Statistics	2 B	CMD	2 B
Phasor angle	4 B	application ID	2 B
Frequency	4 B	sub-query ID	2 B
PMU ID	2 B	phasor name	16 B
application ID	2 B	GPS time	8 B
sub-query ID	2 B	line slope	4 B
		upper and lower bounds	8 B
Total	32 Bytes	Total	58 Bytes

2.6.5 *Creating the Data Model*

The messages transferred in DEFT for BAM have the following steps:

(1) The *Model Creation* at the SPDC involves calculating values for β_s, β_d, Δf_{avg} of (2.7) using the data from both the LPDCs.
(2) The SPDC then disseminates this model to both the LPDCs.
(3) If any LPDC detects a *local violation* given data at t, the corresponding data are sent to the SPDC.
(4) The SPDC then requests for data from other LPDC to check for global violation, i.e., the violation of (2.1).
(5) If SPDC does not detect any *global violation*, it updates the model using the new phase angle and frequency values and sends to both the LPDCs.
(6) If SPDC detects a *global violation*, the operator is alerted to the possibility of stability violation whose phase angle difference is being monitored.

Consider SPDC monitoring an angle difference query, $|\theta_t^s - \theta_t^d| \leq 20$. At time $t = 0$, let the initial values of θ^s and θ^d be 50 and 40 degrees, respectively. To create models at LPDC, SPDC obtains the angle and frequency values from both LPDCs and calculates the local monitoring conditions as given in (2.8). These conditions are then disseminated to both LPDCs to monitor the individual phasor angles. The *Model Creation* requires a total of 90 bytes of data transfer, out of which 32 bytes are from LPDC to SPDC as shown in Table 2.5. At time $t = t_1$, let θ^s and θ^d be 70 and 60 degrees respectively. Also, let the local thresholds for $\theta_0^s - \beta_s$ and $\theta_0^s - \beta_s$ be 75 and 65, respectively. It can be seen from Figure 2.7 that though the actual angle difference is within θ_{Th}^{sd}, θ^s does not lie within its local conditions.

Table 2.6   Data transfer example in BAM.

	Data Transfer between LPDCs and SPDC (bytes)
Model Creation	$90(32 + 58)$
Local violation at one LPDC	$226(32 + 46 + 32 + 2 \times 58)$
Global violation	$110(32 + 46 + 32)$

This leads to a message transfer of 32 bytes, with θ^s and f^s values at time $t = t_1$ being sent from LPDC to SPDC. SPDC then pulls θ^d value from other LPDC corresponding to the timestamp for which θ^s experienced a local violation. This leads to a transfer of two messages between LPDC and SPDC. The request message from SPDC is 46 bytes, and data message from LPDC is 32 bytes. As SPDC detects no angle difference violation, it creates a new model with, possibly, modified local thresholds. The SPDC uses the newly obtained values of θ^s, θ^d, f_s and f_d to modify the local thresholds, and these modified queries are disseminated to individual LPDCs (58 bytes each). Thus, a local violation costs 226 bytes of data transfer. If there was a global violation, there is no model recreation. Thus, in the case of global violation, 110 bytes of data transfer take place between LPDCs to SPDC. Table 2.6 summarizes the data transfer for each of the cases discussed above.

Implementation of DEFT: In DEFT, the time complexity to execute a sub-query is different for different applications. Hence, during the parsing of a PMU packet at LPDC, the application relevant PMU data are processed by an independent thread and then the partially computed sub-query result is packaged and sent to the SPDC. The time-aligning in DEFT is required to temporarily store data — to be retrieved later when local violations occur at peer LPDC for the BAM application. At SPDC, the data packets with partially computed results of applications are parsed and dispatched to the application, to get the final computed results.

Comparison of BAM: DEFT versus CEUT: First, observed latencies for the simulation experiments are discussed and then the processing latency at PDCs of Indian electric grid are estimated.

Estimating latencies at *local* and *super* Phasor Data Concentrators in DEFT: In DEFT, if the LPDC receives the data packets from all the N PMUs at the same time and the data packets relevant for the sub-query of the application are then parsed, the processing latency at LPDC comprises $D^{parse}(K, N)$. The worst case processing latency at LPDC and

SPDC in DEFT is:

$$D_{deft}^{lpdc} = D^{parse}(K, N) + D^{app} \tag{2.9}$$

$$D_{deft}^{spdc} = D^{app} \tag{2.10}$$

where, D^{app} is different at LPDC and SPDC for DEFT. For DEFT, any PDC need not collect and forward the raw data from lower level nodes. Hence, there is no D^{agg} component in their corresponding expressions. However, processing required to send data to higher level nodes is specific to the application, hence, included in D^{app}.

2.6.5.1 Latency for BAM from PMU to SPDC

In this section, simulation results for the test bed are presented. We measure the end-to-end latency from simulated PMU data generation to application specific calculation at SPDC. Figure 2.8 shows the latency for CEUT and DEFT. The benefits of DEFT are evident from the figure. In DEFT, each of the 14 difference angle queries monitored at SPDC corresponded to a model comprising of local conditions that were monitored at both the LPDCs. As the PMU data was sent at 50 packets per second, the number of data points in the simulation duration (30 sec.) for the 14 sub-queries monitored at each LPDC were $50 \times 30 \times 14 = 21000$.

It was observed that, during the *fault duration* of 35 millisec, only twelve global violations occurred whereas very few local violations (1185) occurred at other times. Thus, out of 42000 data points from both the LPDCs, the total number of violations reported were only 1197. Out of twelve global violations, six occurred at time 0.1 sec and rest at 0.13 sec.

The total number of global and local violations accounted for 3% of the entire simulation time whereas the rest 97% of the time all the PMU data were filtered at LPDCs.

Figure 2.8 shows that for this 97% of the time, the latency of BAM with DEFT was just 2.24 millisec as the query execution stopped at LPDC. This latency (D_{deft}^{lpdc}) is mainly due to D^{parse} at LPDC. For the rest 3% of the time, the latency for BAM with DEFT was 8.28 millisec as SPDC pulls data from peer LPDCs for each local violation leading to 32 kb of data transfer for the entire period of simulation. Out of 8.28 millisec, $D_{deft}^{spdc} + D_{deft}^{lpdc}$ accounted for a delay of 5.87 millisec while the rest was due to network delay. This corresponds to 87% and 55% savings in average

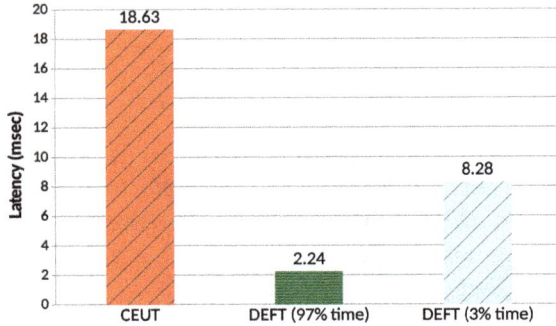

Fig. 2.8   Latency reduction for BAM when run alone. [Khandeparkar *et al.* (2017)]

latency over CEUT when query result was delivered at LPDC (97% of the time) and SPDC (3% of the time), respectively. Even though the query result was delivered to SPDC 3% of the time, the percentage drop is still considerable.

This 3% of time data dissemination from LPDC to SPDC leads to 21 Kbps of bandwidth required with DEFT as opposed to 6.3 Mbps with CEUT.

2.6.5.2   *Estimated Processing Latency for BAM with DEFT on the Indian Electric Grid*

We consider that the angle difference between the substations of Gujarat and Maharashtra states is monitored at Western region PDC. The sub-queries' conditions are then monitored at their respective state PDCs. For DEFT algorithm, two scenarios can occur.

Scenario -1: When the grid is steady, there are no local violations, and the query executes in steps at LPDC. Hence,

$$D_{deft}^{spdc} = 0.$$

At LPDC, the processing latency for the sub-query (D^{app}) is minimal (a few micro secs) and is ignored. So,

$$D^{app} \approx 0.$$

Thus, the estimated processing latency for BAM comprises of parsing time at LPDC

$$D_{deft}^{lpdc} = D^{parse}(K, N).$$

Scenario -2: Local violation at LPDC. For local violation at any LPDC, SPDC pulls data from other LPDCs. Thus, the processing of application at SPDC may be delayed due to queuing of the data request at peer LPDC when it is busy parsing data packets sent by PMUs. As D^{app} for BAM at SPDC is negligible, the processing latency at SPDC is only due to parsing at peer LPDC. So,

$$D_{deft}^{spdc} = D^{parse}(K, N).$$

Impact of large data size: The number of transmission lines per substation is expected to grow soon. The PMUs located at these substations would be required to measure the current (current phasors) of each transmission line. Thus, the data packet size per PMU would increase with an increase in transmission lines at the substation. Furthermore, as per IEEE C37.118.2-2011 standard [IEEE Std C37.118.2], there is no limit on the number of phasors that a PMU can measure. For the Indian grid, there are substation PMUs having more than 2.5 KB data packet [PGCIL (2012)]. In this context, the impact of large data sizes for BAM with CEUT and DEFT in the Indian grid is evaluated.

For PMU data of 2.5 KB, the estimated processing latency for BAM

- without local violation:

$$D_{deft}^{lpdc} = D^{parse}(K, N) = \frac{141.1 * 83}{1000} = 11.7 \text{ millisec.}$$

$$D_{deft}^{spdc} = D^{app} = 0.$$

  Therefore, processing latency without local violation = 11.7 millisec.

- with local violation:

$$D_{deft}^{lpdc} = D^{parse}(K, N) = 11.7 \text{ millisec.}$$

  Also,

$$D_{deft}^{spdc} = D^{parse}(K, N) = 11.7 \text{ millisec.}$$

  Therefore, processing latency with local violation = 23.4 millisec.

The query execution time for BAM at Western region PDC with CEUT is very small (a few micro secs) and hence, ignored. The estimated processing latency at PDCs — this can be calculated for Maharashtra region using Equation (2.4).

$$D_{ceut}^{lpdc} = (141.1 * 83 + 43.6 * 2.5 * 83)/1000 = 20.8 \text{ millisec.}$$

Note that D^{parse} time corresponding to 2.5KB data per PMU is $564.4/4 = 141.1$ is used in the above calculation. This is because the simulator used a quad-core processor as mentioned in Table 2.3.

Considering D^{app} as negligible, the D_{ceut}^{spdc} for Western region can be calculated using Equation (2.5).

$$D_{ceut}^{spdc} = 59.7 * 83 * 2.5/1000 = 12.4 \text{ millisec.}$$

Hence, the processing latency from Maharashtra to Western region PDC is

$$D_{ceut}^{lpdc} + D_{ceut}^{spdc} = 33.2 \text{ millisec.}$$

Thus, BAM with DEFT has 30% and 65% reduction in latency for the query result delivered at SPDC and LPDC, respectively.

In summary, BAM with DEFT achieved significant reductions in latency both in the simulation environment and in the Indian electric grid. Also, in the simulation, DEFT required lesser bandwidth compared to CEUT as data was sent from LPDC to SPDC for only 3% of entire simulation time.

2.6.6   Handling Concurrent Applications

The priority of executing the applications at a PDC depends on the time-critical nature of the application. For example, as given in Table 2.1, BAM has a higher execution priority over MCGG. This is because, in BAM, control action needs to be taken fast enough to prevent propagation of instability throughout the system. PSSE has the lowest priority. Though DEFT exploits the application semantics for data dissemination, LPDCs and SPDCs treat all the application packets equally. This means the processing of data packets relevant to low priority applications would not be preempted even when data packet relevant to a high priority application arrives at an LPDC. In DEFT that considers applications priority, PDCs preempt processing of data packets relevant to low priority applications when packets for high priority applications arrive, thus preserving the order of execution of applications based on their time criticality.

2.6.6.1   DEFT with Priority Cognizance of Applications

In DEFT with priority based execution of concurrent applications, there are separate threads to parse, time-align and perform application specific computations for each PMU packet received. Each of these threads is associated with pre-defined static priorities based on the priority of the applications. This differentiates the processing of more important data packets from other less important data packets when the prior ones are queued at the PDC. Further, if the same data is used by multiple applications, the threads performing application-specific computations are associated with priority values in the decreasing order of their criticality.

All the data dissemination techniques were implemented over a Linux platform. The Linux scheduler decides which runnable thread is to be executed by the CPU based on real-time scheduling policy (SCHED_FIFO, SCHED_RR, etc.) and scheduling priority of the thread [Michael Kerrisk (2014)]. The priorities are in the range 1–99, with 1 being the lowest and 99 the highest priority. SCHED_FIFO scheduling policy was used for all threads with appropriate scheduling priorities based on the criticality of the task performed by the thread. As given in Table 2.1, threads that parse PMU data packets corresponding to BAM have a high priority (95), MCGG has medium priority (90), and SE has low priority (85). The threads corresponding to the time-align operation has the lowest priority (80) among all other tasks.

Henceforth, we will refer to DEFT that does not consider application priorities under concurrent execution as plain-*DEFT, while the one that considers the application priorities as DEFT.*

2.6.6.2 *Latency from PMU to SPDC*

Figure 2.9(a) shows the average latency with *plain*-DEFT and DEFT for BAM when run singly and concurrently with other applications. We see that unlike *plain*-DEFT, with DEFT, the average latency remained almost the same when BAM was run singly and concurrently with other applications. As DEFT executed BAM with the highest priority, having other applications executing concurrently does not have much impact. The marginal increase in latency of BAM with DEFT from 8.35 millisec to 8.45 millisec was due to a thread context switching time during preemption of low priority threads. Figure 2.9(b) shows the average latency of MCGG with DEFT to be higher when run with BAM compared to *plain*-DEFT. This is because, in DEFT, threads processing MCGG relevant data got preempted by threads that process BAM relevant data. The latency of MCGG with DEFT slightly reduced when run with PSSE unlike when run with BAM, as MCGG has higher execution priority compared to PSSE.

Figure 2.9(c) shows that the average latency for PSSE with DEFT was more than *plain*-DEFT as PSSE has the lowest priority among all the applications, in DEFT, the threads processing of packets relevant to PSSE would always be preempted by other threads.

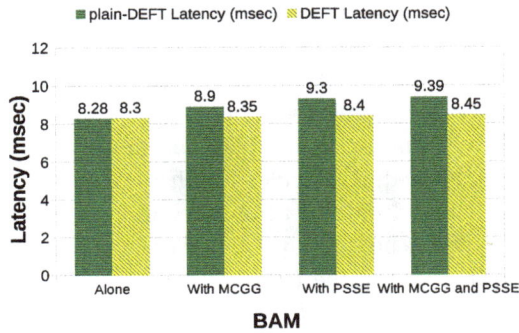

(a) BAM: singly and concurrently

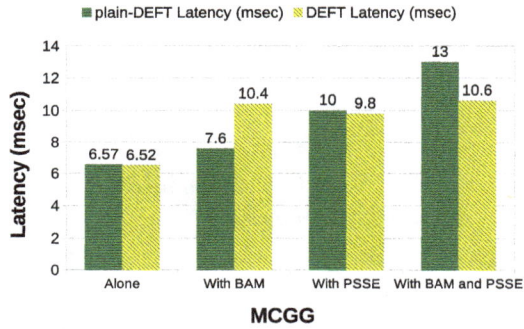

(b) MCGG: singly and concurrently

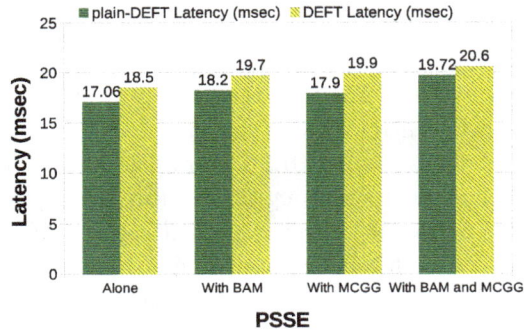

(c) PSSE: singly and concurrently

Fig. 2.9   Average latency: running singly and concurrently with *plain*-DEFT and DEFT. [Khandeparkar *et al.* (2017)]

From Figure 2.9, it can be summarized that as DEFT takes cognizance priority of the applications, it is more suitable for time-critical applications.

2.6.7 *Making All Data Available at SPDC*

Our discussion so far has been confined towards the application semantics-aware PMU data dissemination. The SPDC, however, may require all PMU data from downstream LPDCs/PMUs to perform post-analysis of certain events such as load trips, generator trips, etc. This system wide data from all PMUs can reveal insights. Further, sending raw data from PMUs provides an additional level of reliability to loss of packets in the network. Hence, existing DEFT approach to *send all PMU data to SPDC* is extended. Two methods are proposed viz., i) DEFT+CEUT and ii) DEFT+CEUT-*direct* as discussed in Section 2.5.2.3.

In the first method, just like application specific data, all the PMU data is time-aligned at LPDC, a combined packet is created, and disseminated to SPDC. This raw PMU data is typically sent at the lowest priority. CEUT-*direct* has already been discussed in Section 2.6.3. In this method, PMUs multicast data to both LPDC and SPDC. The first method requires an additional combined packet creation operation at LPDC whereas, the second method requires additional parsing of PMU data packets at SPDC.

2.6.7.1 *Latency from PMU to SPDC for Applications with All the Data Dissemination Techniques*

Figure 2.10 shows the latency for individual applications with all the data dissemination techniques on the simulation test-bed. *Plain*-DEFT and DEFT have already been compared earlier. With DEFT+CEUT, the latency for all applications was almost the same as the latency with DEFT. This implies that sending all PMU data to SPDC with the lowest priority does not significantly affect the latency of individual applications. However, with DEFT+CEUT-*direct*, the latency of individual applications slightly increased in comparison to DEFT+CEUT. This is because, before the application specific data from LPDCs could reach SPDC, PMU data packets from PMUs had reached SPDC. As SPDC was busy processing a total of 662 data packets from PMUs of Eastern and Western electric grid, even though the packets with partially computed results had reached the SPDC, their processing was delayed due to the preempting threads processing PMU data packets.

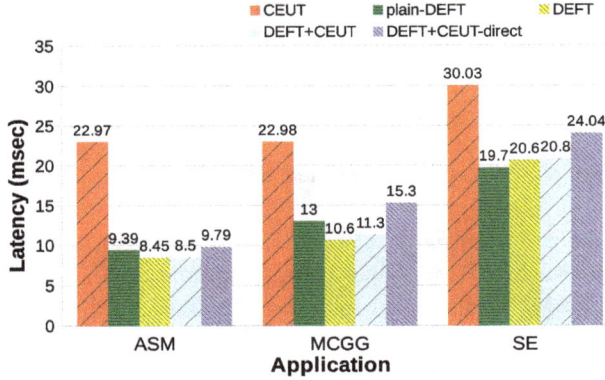

Fig. 2.10   A comparison of latency for all the data dissemination techniques. [Khande-parkar *et al.* (2017)]

2.6.7.2   *Latency from PMU to SPDC during Loss of Filtered Data Packets for BAM from LPDC to SPDC with DEFT+CEUT*

Let us now evaluate the performance of DEFT+CEUT when the filtered data packets sent from a LPDC for BAM application is lost. Under these circumstances, the applications (at SPDC) can continue to execute the difference angle query using the angular information extracted from the raw PMU data that is sent by both the LPDCs at the lowest priority. We also consider here MCGG and PSSE applications concurrently running with the BAM application. Without any loss of the application specific data packets, the PMU to SPDC end-to-end observed latency for BAM as shown in Figure 2.10 was 8.5 millisec.

When a filtered packet for BAM from any one LPDC is lost, SPDC needs to wait for the arrival of the aggregated data packet sent by each LPDC with CEUT. Before sending the aggregated data packet from an LPDC, the sub-queries of MCGG and PSSE are executed with DEFT. Hence, processing latency at LPDC comprised of observed D_{ceut}^{lpdc} (16.3 millisec) to parse and aggregate data packets from all PMUs including thread pre-emption overheads and observed D^{app} for the sub-queries of BAM, MCGG and PSSE. The sum total of D^{app} for BAM, MCGG and PSSE was 3.4 millisec.

At SPDC, as the execution of BAM, whose filtered data packet was lost, can be done by extracting phase angle information from the aggregated data packet, the processing latency was dominated by D^{parse} (2.0 millisec).

Thus, PMU-SPDC end-to-end latency for BAM during the loss of filtered BAM packet was 24.3 millisec.

This average latency with DEFT+CEUT is slightly greater than that of BAM with CEUT (22.97 millisec) as shown in Figure 2.10. However, we note here that as the aggregated raw PMU data is also available at SPDC in DEFT+CEUT, it is suitable under loss of in-network filtered data packets of any application sent from LPDCs to SPDC.

2.7   Summary and Takeaways

PMUs allow the state of the power system — the voltage phasor of system buses and current phasors of all incident transmission lines — to be directly measured and in some cases inferred at neighboring buses and lines [NASPI (2009)].

Typical PMU data measured at a substation (generation or transmission) has three voltage phasors, three current phasors, one frequency value, a GPS time-stamp, and other analog and digital values [IEEE Std C37.118.2]. In a smart grid, the signals are typically sampled and communicated at high rates (several 10s or 100s of times per second), which demands augmentation, if not replacement of the conventional supervisory control and data acquisition (SCADA) system in which measurements are available at best every 4 second [Bobba *et al.* (2012)] from the RTUs. A typical PMU data packet size is 100 bytes, but larger packet sizes are expected to be the norm soon, as more and more grid parameters get included for monitoring [PGCIL (2012); IEEE Std C37.118.2]. Thus, with PMUs having 90 phasors and 45 analog values at 100 Hz of sampling frequency, data from a single PMU could exceed 700 kbps. For India's grid, a total of 1600 PMUs is envisaged, making the total data size exceed 1 Gbps. These PMU measurements are synchronized using a GPS clock, enabling a consistent snapshot of the system. These provide the data necessary to assess grid performance and enhance the ability to control system operations and management.

Various analytical tools and algorithms are used to aggregate and analyze the PMU data from different geographical locations [Phadke (1993)]. Phasor Data Concentrators (PDCs) at one or more levels aggregate and integrate the PMU data. Lower level PDCs (LPDCs) aggregate data from PMUs that are located at different places, time align the data, and send the aggregated data to a higher level or super PDCs (SPDCs). In comparison, our technique uses *application semantics* to perform in-network query processing for smart grid applications. Thus, it can be used for exact monitoring and not just probabilistic monitoring.

Data traffic reduction using filtering at the data sources is proposed in [Gupta *et al.* (2010); Olston *et al.* (2003)]. These filtering conditions are derived from sufficient conditions for providing *no-false-negative* guarantees for threshold queries. Authors of [Gupta and Ramamritham (2012)] use data aggregation across distributed sources. These techniques perform in-network query processing on sensor values. In comparison, we use data modeling and in-network processing to monitor the grid efficiently.

Various techniques for QoS guaranteed delivery of data have been studied in the literature. Gridstat [Bakken *et al.* (2011)] discusses the implementation of a data dissemination network with delivery guarantees in a smart grid. Our approach optimizes data dissemination and is orthogonal to Gridstat. Thus, this solution can be used with Gridstat or any other data delivery mechanism. It also retains the existing client-server framework for PMU-PDC communication that is currently adopted by the power system domain — unlike Gridstat, which is a new middleware paradigm. The stream computing based approach for optimizing the data transmission in [Hazra *et al.* (2011)] makes way for an easy development of applications and allows for graceful degradation during times of overload.

Chapter 3

Energy Management Systems for Modern Buildings

Building Energy Management Systems consist of monitoring and control components — cyber (i.e., smart sensors, digital electronic hardware, software) components tightly integrated with physical components (i.e., the building itself and loads, appliances and actuators within the building). Modern buildings have significant monitoring, computational and control capabilities embedded in them, where the monitoring is done by analyzing the data from sensors and the various aspects of the building's operation are controlled automatically. Such buildings are also referred to as *smart buildings* or *smart homes* due to their embedded intelligence and automation capabilities. Modern buildings implement the sense, analyze, respond with steps of the SMART approach, outlined in Chapter 1.

To illustrate the capabilities of today's smart buildings and homes, consider the following examples. Office buildings often employ automated lighting control where the lights in a room are controlled by motion sensors — they turn OFF when no motion is sensed in the room to save energy and turn ON automatically when someone enters the room. Similarly, office buildings employ automated shading control on windows where curtains, shades or blinds are motorized and controlled automatically to allow natural light whenever possible and also to limit excessive sunlight from causing unneeded heat gain. Office buildings also employ smart HVAC controls where the heating and cooling is automated based on the occupancy levels in the building — for example, the heating or cooling may be turned down in the night when the office building is unoccupied and may be automatically turned ON to pre-cool or pre-heat the building before people arrive in the morning. In residential settings, smart homes can automatically unlock the front door when the system senses a resident (with security check on identity) outside the door, or can turn OFF the air-conditioning in a room when it senses a window being opened in the room. This chapter and the following ones describe the various technologies and techniques needed to implement such smart automation.

Today's buildings can be broadly categorized into two types: residential and commercial. Residential buildings include single family homes (bungalows), apartment buildings, condominiums, townhouses, and dormitories. Commercial buildings are more varied in nature depending on their purpose and include office buildings, shopping complexes, malls, warehouses, hotels, hospitals, airport terminals, industrial complexes, among others. Depending on whether a building is residential or commercial in nature, different technologies have been used for implementing the sense, analyze, and respond CPS functions for the building. Sections 3.1 and 3.2 provide an overview of these technologies for commercial and residential buildings, respectively.

3.1   Commercial Buildings

Most large commercial buildings have automated monitoring and control capabilities that are implemented through a *Building Management System (BMS)*.

> ### Building Management System (BMS)
>
> A building management system (BMS) is a computer-based control system that manages and monitors the building's mechanical and electrical equipment such as air-conditioning, ventilation, lighting, power systems, fire systems, and security systems.
>
> A typical building management system consists of three key components: (i) sensors and actuators that are deployed to monitor and control various aspects of a building's operation, (ii) the BMS control node that runs software to monitor data from sensors and remotely send control signals to various actuators while also providing automation capabilities; (iii) a communication protocol to connect sensors and actuators and enable communication to and from the control node.

Figure 3.1 depicts a building equipped with a Building Management system comprising these components. Typically a BMS is used to control

Fig. 3.1   An illustration of a residential building with a building management system.

various aspects of a building's operation including:

- Lighting control, where lights can be remotely turned ON or OFF based on a preset schedule (e.g., in hallways) or based on occupancy,
- Power control, where electricity usage in various zones of the building can be monitored and backup systems such as UPS can be controlled,
- Heating, Ventilation and Air-Conditioning (HVAC) control, where heating and cooling in various parts of the building can be independently monitored and controlled,
- Plumbing control, where water usage in various parts of the building can be monitored,
- Fire and Safety, where smoke detectors are used to detect fire, generate alarms automatically and remotely actuate safety systems such as sprinklers,
- Security systems, including CCTV cameras to monitor the security of the building and access control that secure doors in various parts of the building, and
- Systems to monitor elevators, escalators and other systems.

Certain variants of BMS are referred to as Building Energy Management System (BEMS) as they help manage energy consuming devices and equipment in the building, such as heating, ventilation and air conditioning (HVAC) systems and lighting, which contribute to over 70% of the building's energy needs. Since a BMS provides automation capabilities where many aspects of a building operations such as lighting, heating, cooling and access control can be automated, they are sometimes also referred to as Building Automation System (BAS). Such automation can take many forms. Lighting automation, for example, can be closed loop or remote controlled. In case of closed loop control, lights in a room are connected to motion sensors, which turn the lights ON whenever someone enters the room and turn them OFF when the room is unoccupied, thereby preventing energy wastage. Strictly speaking, closed-loop lighting control does not require a central node such as BMS since lights are driven by local sensors in a "closed loop". However, a BMS is useful for remote control of lighting where lights in common areas of the building such as hallways or atria can be turned ON or OFF based on a pre-determined time schedule and often in combination with illumination levels. Since HVAC systems consume more than half of the building's total energy usage, a BMS offers sophisticated control of the HVAC system to optimize energy use. This includes fine-grain control of the HVAC system where heating and cooling can be independently controlled for each zone, floor or even individual rooms. In cases where the centralized HVAC system uses ducts to provide air flow and climate control, a BMS can enable control of airflow (e.g., variable air volume or VAV system) delivered through each vent. Another common feature is the ability to use pre-set schedules for different days of the week, including weekends, that determine when the heating and cooling come on in different parts of the building — depending on when people arrive or depart. Such automation capabilities are at the heart of intelligent energy management using computational methods that we describe in later chapters of this book.

Communication Protocols in BMS

A building management system needs to employ a communication protocol to network the sensors and actuators under its control. In commercial buildings, most common protocols are wired in nature, unlike residential settings where wireless protocols are more common. A number of communication protocols have been developed over the years for building management and building automation. These include Modbus, BACnet, and LonWorks among others.

Modbus is a serial communication protocol that has become the de facto standard for connecting electronic devices in industrial environments. Modbus was developed by Schneider Electric and is now managed by the Modbus organization which has made it available in open and royalty free form [Modbus; Modbus Protocol; Modbus TCP/IP]. Modbus works in a master-slave architecture, with one centralized master and up to 247 slaves.

The MODBUS protocol defines a simple Protocol Data Unit (PDU), which is independent of the underlying communication layers. The mapping of MODBUS protocol on specific bus or network can introduce some additional fields to generate the application data unit (ADU). In case of RS-232/RS-485 communication, one byte header for the device ID and 2 byte CRC fields at the end of the PDU create the original Modbus-RTU message (ADU). For Modbus on Ethernet TCP/IP, a dedicated 7 byte header called MBAP header (MODBUS Application Protocol header) is added and there is no field for CRC at the end of the ADU. The most common usage of Modbus is in supervisory control and data acquisition (SCADA) systems where a number of remote control-units are connected to a centralised SCADA station. SCADA itself is a control system architecture for connecting industrial controllers and managing them remotely.

BACnet is a protocol designed for Building Automation and Control (BAC) that defines data communication services and protocols for computer equipment used for monitoring and control of heating, ventilation, air-conditioning and refrigeration and other building systems [ISO 16484]. Unlike ModBus, which was designed for general industrial use, BACnet specifically targets building management and is an ISO standard (ISO 16484-5:2003). It is designed to operate over many data link layers and MAC protocols such as EtherNet, BACnet/IP and RS-232 serial communication. BACNet defines an abstract, object-oriented representation of data as well as a set of services, such as device and object discovery, that enables a control node and building equipment to interact with one another.

LonWorks is a networking protocol specifically designed to address control applications and is typically used for automating lighting and HVAC equipment in buildings. LonWorks is both an ANSI and an ISO standard (ISO 14908) and can operate over twisted pair cables, powerlines, fiber optic cable, and RF. It can also be used as a data link layer for BACnet and is currently in use in applications such as electric metering, lighting, fire detection and HVAC systems.

3.2 Residential Buildings

Unlike commercial buildings, residential buildings have typically not been equipped with sophisticated building management systems. However this is changing with the rise of the Internet of Things and today a number of IOT-enabled products are available for smart building automation in residential settings. The typical architecture for a smart home includes five components:

- Sensors, smart IoT devices and smart appliances that monitor and control various aspects of the home,
- A smart home hub which acts as a control node that interfaces with sensors and smart devices and also with the end user,
- One or more communication protocols, typically wireless, that are used to network the smart home hub with the sensors and smart appliances,
- A cloud service that allows remote access to the hub and the smart devices from outside the home, and
- Mobile phone apps that allow end-users to interact with the hub and all the smart devices within the home.

IoT-based commercial products are now available for smart home automation and energy management including:

- Smart switches and outlets, which enable remote control of lights and any device plugged into an outlet
- Smart light bulbs, which enables remote on, off, and brightness control of a light bulb through wireless commands
- Smart thermostats, which allow HVAC equipment within a home to be remotely controlled and programmed,
- Smart appliances, such as smart fridge, smart washing machine and others, with provisions for remote monitoring and control,
- Smart electric meters, which allow end-users to directly monitor their electricity usage through sensors that are installed within the circuit breaker panel or fuse box of the home,
- Smart locks, which allow door locks on entry doors to be remotely locked or unlocked,
- Door and window sensors, which sense where the door or window is open or closed,
- Occupancy sensors, which include PIR motion sensors to detect motion within a room or counting door sensors that track when someone enters or leaves the room, and
- Security camera, which can monitor an area and stream video feeds to a remote server (e.g., when motion is sensed).

Figures 3.2 and 3.3 illustrate two commercially available IoT products, namely Wemo smart outlet and Nest smart thermostat, for smart homes.

Fig. 3.2 Wemo Outlet, a commercially available smart outlet, provide remote on-off capability as well as the ability to monitor the power usage of plugged in devices.

Fig. 3.3 Nest Learning Thermostat, a commercially available smart thermostat, that can automatically determine a heating and cooling schedule for the home based on observed occupancy patterns.

Since a variety of products, made by different manufacturers are available, each exposing their own interface, the user will need to interact with each device using its native interface. This can become cumbersome in homes with a large number of networked sensors and smart devices. Smart home hubs address this issue by interfacing with the myriad sensors and devices within a home and exposing a single unified interface to the end user. The smart home hub plays the same role as the control node of a BMS in commercial settings — it provides monitoring capabilities with sensors, provide remote control as well as automation. Automation is typically provided through a rules engine, which allows users to specify an action whenever a trigger event is observed (e.g., if motion is sensed in a room, turn the lights on). Many hubs provide a custom rules engine, while others use Internet services such as IFTTT [IFTTT] for this purpose. Since a hub needs to interface with a diverse set of smart devices, they tend to support multiple communication protocols and multiple interfaces to ensure broad interoperability. Examples of popular commercially available smart home hubs include Wink, Iris, SmartThings, MiCasa, and Apple Homekit (which runs the hub software on apple devices such as a Mac computer or Apple TV).

3.3   Sensing Facets of a Building Meaningfully

Since a BMS in a commercial building or a smart hub in a residential building manages a range of tasks such as building health monitoring, occupancy tracking, maintaining comfort, HVAC control, and energy management, it is important to deploy various sensors, actuators and smart devices at appropriate locations to enable these tasks. Thus, regardless of whether such instrumentation is done at construction time or as a retrofit to an existing building, *which sensors and actuators to deploy and where* is an important design problem.

Perhaps the most important aspect of a working building, of interest to a BMS, is the power consumption: How much power is consumed at any point in time, and also its historic profile. If the BMS can also determine the appliances that are ON at a certain point in time, then it can take steps to reduce or optimize consumption.

At the outset, we identify the electrical loads (refer to Section 7.3.1 for a discussion on non-intrusive monitoring of different types of electrical loads — NILM) causing the prevailing power usage. Then we look into other facets or properties of the building that we would like to track and discuss.

A smart building needs to monitor the *states* of the building in terms of i) power consumption, ii) possible wastage of power, occupancy, iii) thermal comfort level, etc. This requires a wide range of sensors so that the building management system can monitor and carry out its tasks of reducing and optimizing power consumption, monitoring the health of the building appliances, maintaining quality of the atmosphere in the building and tracking occupants in various parts of the building (useful for purposes such as building safety and emergency evacuation), etc.

Since sensing infrastructure or needs can vary greatly in buildings, it is not feasible to deploy sensors "everywhere." Doing so can greatly increase

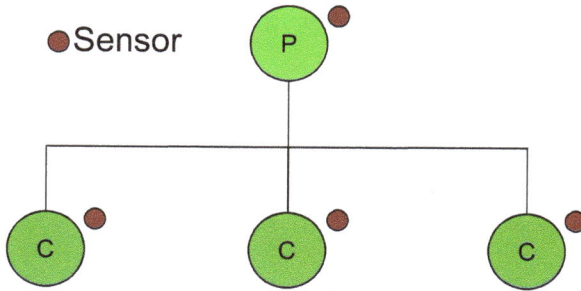

Fig. 3.4   Redundant sensing.

the cost and payback period, while also impacting building aesthetics and causing user inconvenience. Furthermore, this will lead to a great deal of redundancy in the sensing infrastructure.

First, redundancy can be inherent due to the structure of the sensing problem. As an example, consider Figure 3.4 where sensors are placed at every node of a tree, where the leaf nodes (marked as child nodes C) represent individual power outlets, while the parent node P represents the circuit breaker that enables power supply to these outlets. It can be observed that the sensor on at least one of the nodes is redundant. Thus, if we measure power usage at each outlet, the total usage can be obtained as the sum of the power used by the individual outlets. Similarly the power usage of an outlet C can be computed from that flowing through the common breaker and the other two outlets.

In some cases, there may be similarities in the power usage of the leaf nodes. For instance, the same type of load may be plugged into all three, or the leaf node may represent the power usage of an office room and all office rooms may be similarly equipped. In this case, *sampling* one or a small number of leaf nodes may be sufficient and the power usage of the other leaf nodes can be estimated from these samples.

Second, there may be redundancy from different types of sensors. For example, one type of sensor deployed in an area may reveal other types of information, typically obtained using sensors dedicated for that purpose. As an example, consider motion-activated lighting. Since the motion sensor turns lights on or off based on motion (occupancy), there is no need to monitor separately the power usage of the lights, since it can be directly computed based on whether there was motion in the room. The reverse is also true — monitoring power usage in an area can also reveal occupancy, since more power events (e.g., manually turning lights on or off) indicates the presence of users in that area.

Third, a building may include soft sensors that are part of the IT infrastructure that may reveal certain types of information, making hard sensors redundant. For example, WiFi activity at an access point or active network traffic from a desktop reveals the presence of humans in the vicinity, and can be used as soft sensors for tracking occupancy in that part of the building.

Fourth, certain types of data can be inferred using more sophisticated machine learning and inference algorithms that take inputs from other hard (physical) or soft sensors. Load disaggregation algorithms that infer the power usage of individual loads from aggregate measurements of total power usage is an example of such type of redundancy.

Fifth, depending on the application goals, it may be possible to exploit redundancy in deciding the number of sensors needed to be deployed. For example, if an application needs to monitor whether a home is occupied, one approach is to deploy motion sensors in every room. Another approach is to deploy door sensors that track how many people enter and leave the home, maintaining a count of how many people are inside. If the application only needs to know whether the building is occupied, the door sensors are adequate for this task and room-specific motion sensors can be eliminated. However, if the application also needs to know spatial occupancy details of which rooms are occupied, the door sensor is inadequate and each room needs to be instrumented with motion sensor. Thus, the same problem of occupancy monitoring may require different levels of instrumentation depending on the goals of the higher level application.

Thus, it is possible to exploit these redundancies to vastly reduce the number and types of sensors and actuators that need to be deployed without sacrificing quality and resolution of sensor data while meeting the application-specific goals. To answer our previous question of what sensors to deploy in a building and where, we must consider the following aspects:

• Different parts of a building have similarities in their consumption profile.
• Redundancy inherent in the hierarchical nature of the building sensing infrastructure.
• Observations made in the context of one facet to infer another facet of interest.
• Minimized need for physical sensing due to the availability of soft sensors.
• Goals of the specific energy management tasks (i.e., application goals).

3.3.1   Terms Related to Sensing within a Building

In this section, we introduce and define the common terminology used in tandem with the topic of meaningful sensing.

Facet — A facet is defined as the factor that influences or represents the state or behaviour of the building. For example, power consumption might denote the state of the building as how much power is being consumed by the building. Occupancy might influence the power consumption, thus influencing the state of the building. Similarly temperature and humidity can be examples of other facets.

Observation — An observation represents the value of a facet. Obtaining an observation is tantamount to observing a value.

Observability — Observability of the facet is defined as the ability to observe the value of the facet. Observability of a facet at a node implies that the system is able to obtain the observations of the facet at the node. The notion of observability has been introduced to monitor observations of different factors influencing the state and behaviour of buildings.

Information about a particular facet of some space can come from physical measurements of certain phenomena or inferred from some information pertaining to (constituents of) that space.

Important facets of interest in the context of buildings are power consumption, occupancy i.e., number of occupants, number of ON appliances and temperature. Let us now look into the typical *Hard or Physical sensors* and the *Soft or Inferred sensors* in a building, which are (can be) used to observe the above facets.

3.3.2   Hard Sensors

Hard sensors are real or physical sensors that are deployed in a building to observe one or more facets. The following sensors are examples of hard sensors that can be deployed within a building.

Temperature and humidity sensors are used to sense the temperature and humidity of the environment.

Smart meters (SM) can provide data on voltage, current, power factor, etc. with high frequency and also allow remote data collection. These facilitate analysis of the energy consumption.

Smart outlets and smart switches can be used to monitor and control individual appliances that are connected to a power outlet or a switch, by turning them on or off through remote commands and by tracking their current state and power usage.

A *Counting Door* (CD) is a plug and play sensor which can keep track of the number of people in a building. A Smart Door described in [Chil Prakash *et al.* (2015)] can additionally keep track of the occupants' identity.

A *Clamp-on meter* is a plug and play device, which is used to measure instantaneous power consumption. Some clamp-on meters also allow remote data collection.

Passive Infrared sensors (PIR) are used for motion detection.

Camera is used to monitor an area.

3.3.3   *Fusion of Hard Sensors for Occupancy Sensing*

Occupancy sensing using hard sensors, is perhaps the most common form of sensing encountered in buildings. There are various kinds of occupancy sensors using different sensing mechanisms, e.g., passive infra-red (PIR), radio frequency Identification (RFID), CO_2 levels, sound detection. We will discuss advantages and limitations of these occupancy sensing techniques in this section. In addition, we will also discuss how personal Wi-Fi enabled devices, LAN activity, etc., can also be utilized for occupancy detection and tracking.

Most occupancy detectors use Passive Infrared (PIR) sensors. These detectors are cheap and easily available in the market. But, erroneously infer the lack of movement for a pre-defined duration of time as non-occupancy, thereby, generating false negatives if occupant(s) become stationary (e.g., reading a book or working on a laptop) inside a space under detection.

Another familiar occupant monitoring system is the Active Radio Frequency Identification (RFID) based system. RFID systems are generally used for access control in buildings, but it also logs the occupant identity which can be used to monitor occupancy. There are two kinds of RFID system — Active and Passive. Active RFID systems, unlike passive RFID systems, do not require a stop and swipe mechanism. They are also known to have high accuracy. But in order to obtain high accuracy users need to carry the tag at all times, they also have to be careful not to keep the tag next to metallic objects and be wary of the speed with which they walk across the RFID reader. Moreover, RFID readers are expensive and generally not deployed in each and every doorway of a building. Hence, this technique is not suitable for room level occupancy.

In case of occupancy counting and directional detection, both PIR and RFID sensors are used.

Direction is identified with the help of twin beam PIR detectors using the sequence of obstruction of the two beams. But, twin beam PIR technique requires a door with certain depth and not suitable for an ordinary door leading straight to a room.

A set of two RFID sensors are required to detect the direction of movement of occupants — one for INbound and another for OUTbound. The advantages and disadvantages of occupancy detection and counting using PIR and RFID sensors are summarized in Table 3.1.

Table 3.1   Occupancy tracking with single type of sensors.

Sensor	Advantages	Disadvantages	Information
Passive Infra Red (PIR)	Cheap; Scalable; RT Response; Requires no user intervention	When users become stationary (e.g., working on PC) room occupancy detected as NIL	Presence of occupants in room and also count in case of twin beam PIR sensors
Radio Frequency Identification (RFID)	Accurate when proper measures are taken; Real Time response; Requires no user intervention	User must carry RFID tag; Tags must not be kept near metallic objects; Accuracy depends on speed of walking [5]	Occupancy Count and identity
Camera	High Accuracy; Requires no user intervention	Requires high bandwidth due to continuous monitoring; Costly;	Occupancy and also count with image processing.

The disadvantages of using a single type of sensor, discussed above, can be overcome by using multiple types of sensors or fusion of sensors. In order to detect direction, a reed switch that actuates on the closing of the door is used in combination with PIR sensor. The sequence of actuation of reed switch and the detection of movement by PIR sensor identifies the occupancy accurately. (The sequences i) reed switch followed by PIR detection and ii) PIR detection followed by reed switch are inferred as entry/presence of a person and exit/absence, respectively.) But, this technique is most suitable for occupancy detection in single user rooms and fails in multi-user environments. When a camera is used in combination with a PIR sensor, we are able to detect occupancy with a high accuracy. While using only a camera can also accomplish this task with relatively high accuracy, its high bandwidth requirement is a downside, as mentioned in Table 3.1.

The fusion of PIR sensors and camera can overcome this disadvantage and it works as follows.

- An array of the required number of PIR sensors are distributed throughout the space in order to ensure that there is no blind spot.
- The camera is triggered by the PIR sensor(s), only to confirm occupancy status detected by PIR sensors. Thereafter, the camera checks for non-occupancy only periodically. Thus the system is reliable and produces relatively less amount of data.
- Data transfer from cameras is done in a scheduled manner in order to counter a possible scenario where cameras in all spaces in the building are triggered simultaneously. This technique also distributes the image processing load on the server required for occupancy detection and counting.

Table 3.2 summarizes the options for occupancy detection by sensor fusion, and their advantages and disadvantages.

Table 3.2   Occupancy tracking with fusion of sensors.

Sensor	Advantages	Disadvantages	Occupancy Information
PIR+Reed[1]	Cheap; Scalable	Fails to detect occupancy in a multi user environment	Presence of occupant in room
PIR+Camera	High accuracy	Use of camera can make it costly; Useful only in selective public spaces due to privacy issue	Presence of occupant in room

3.4   Smart Sensor Suite for Buildings: A Case Study

This case study explains hardware modules that can be deployed in an existing building along with software tools which are open source to reduce the licensing cost.

3.4.1   Hardware Architecture Design

The sensor suite is designed in such a way that a set of dumb hardware modules communicate with each other to make the environment smart, i.e., it takes its own decisions such as whether to turn the Lights and Fans ON/OFF depending upon the occupancy data and temperature data. To realize this, we divide the sensor suite into three general types of nodes:

(1) Controller Nodes
(2) Sensor Nodes
(3) Actuator Nodes

The hardware architecture of each type of nodes is briefly described in the following segments:

3.4.1.1 *Design of the Controller Node*

Controller node can be considered as the *Central Node* in our sensor suite as it handles the entire wireless communication as well as data processing. Therefore, the hardware for this node must be powerful enough to handle multiprocessing and wireless communication. Apart from these, it should have low cost and small hardware footprint, so that it does not affect the aesthetics of the buildings and increases the portability of the node.

To accommodate these properties, we chose Raspberry Pi 3 Model B as our controller node. It has high processing power (A 1.2GHz 64-bit quad-core ARMv8 CPU), inbuilt 802.11n Wireless LAN and Bluetooth Low Energy (which means, no external hardware) with dimensions equivalent to a credit card. These properties make the Raspberry Pi suitable for our controller node hardware.

The raspberry pi runs multiple threads. Some threads initiate the system on wake up/reboot; others listen to the data transmitted by the sensor nodes, a process that captures data and then transmits the decisions such as ON/OFF signals to the actuator nodes over WiFi. The node stores the temperature sensor data in a file and then sends it to the server side over LAN. This is used, for example, for *Thermal Profiling*, explained in the later sections.

As the entire process of collecting data from sensors and transmitting control signals based on that data to the actuators is wireless, this makes the controller node portable and allows the user to place it anywhere in the building as long as it is connected to the same wireless network in which sensor nodes and actuator nodes are connected.

3.4.1.2 *Design of the Sensor Node*

In order to make appropriate control decisions to increase the system's reliability; we should have ample data about the environmental parameters such as temperature, humidity, occupancy. Therefore, we must sense these parameters from different locations of the building. To achieve this, we designed the *Sensor Nodes* which a) accurately measure the above parameters, b) can be deployed in remote locations, c) powered using a battery, d) have a hardware footprint as small as possible. The sensor nodes are divided into two parts a) *Master Sensor Nodes* and b) *Slave Sensor Nodes.*

Table 3.3   Sensor types.

Sensor	Parameter	Interface	Range	Accuracy	Operating Voltage	Active Current	Cost
DS18B20	Temperature	One Wire	$-55°$C to $+125°$C	$±0.5\%$	$+3.0$ to $+5.5$ v	1.5 mA	$3.95
DHT22	Temperature & Humidity	Digital Pin	T:$-40°$C to $+80°$C H:0–100% RH	T:$±0.5\%$ H:$±2\%$	3.3 v to 6 v	2.5 mA	$9.95
PIR (Pololu)	Motion	Digital Pin	7 m & $<120°$	N/A	5 v to 20 v	65 mA	$9.95
PIR (Panasonic)	Motion	Digital Pin	5 m	N/A	2.3 v to 4 v	100μ A	$28

Master Sensor Node acts as a gateway, collecting the data from all the slave sensor nodes and transmitting it to the controller node over WiFi. The hardware architecture of this node uses Arduino Pro Mini as Master Controller, nRF24L01+ RF communication modules to communicate with the slave sensor nodes and ESP8266 WiFi module to transmit data to the controller node. The RF module is connected to the Arduino over *Serial Peripheral Interface* (SPI) whereas EP8266 module communicates with the controller using UART bus. The master sensor node has to be powered using DC Power Adapter.

Slave Sensor Nodes are low energy consuming battery powered nodes which can be deployed across the building to read the different parameters accurately and then send it to the master sensor node. This node consists of Arduino Pro Mini as a slave controller, nRF24L01+ RF communication module, and a sensor. Depending on the application requirement, this sensor could be a temperature sensor or a motion detector sensor.

Dallas Semiconductor's *DS18B20 One Wire* temperature sensor is used to sense the temperature of the area. The advantages of this sensor are that multiple such sensors can be connected to the controller using just single wire bus which reduces the hardware footprint significantly. The slave controller in such nodes will be in sleep mode, and it will wake up periodically to sense and transmit the data. It is smart in the sense that it can set its period depending upon the rate of change of temperature i.e., the higher the rate of change of temperature, the higher the transmission frequency and vice-versa, therefore, providing quality data and also reducing current consumption.

On the other hand, *Passive Infrared (PIR)* sensors are used to detect the motion inside rooms. The slave controller in such node is in sleep mode unless and until PIR detects the motion. The moment PIR gets triggered, it sends an interrupt signal to the slave controller to wake it up from sleep mode. It reads the data from the sensor, transmits it to the master node and then again goes into the sleep mode. The slave sensor nodes consume 30 mA while transmitting data and 0.1 mA while in sleep mode. Therefore, the estimated life of the nodes with 2200 mAH battery at 1-minute data transmitting frequency is almost eight months!

3.4.2   *Communication Protocol*

Communication is an essential aspect of distributed systems. Various types of communication are possible in a system: (a) one to one, (b) one to many, (c) many to one, (d) many to many. Communication protocol suitable for a sensor suite must support all types of communication. Various design requirements are considered while selecting a communication protocol and for comparative analysis of the protocols. Design requirements for a protocol include:

- Support of client-server architecture
  - Sensor, Actuator are analogous to client and controller acts as a server. Client sensor while sensing a phenomenon sends the sensed value to the controller. The controller as a server entity collects the value, processes it and then sends its corresponding commands to the actuator.
- Decoupled entities in the system
  - To increase flexibility and dynamics while building the system it is better to have a decoupled communication between the entities involved because it helps updating and maintaining the system.
- Robustness
  - The protocol should have a mechanism from retransmission and reconnection, after disconnecting, as there are chances of failure at various network nodes.
- Easy to use on various embedded devices
  - Hardware requirement of the protocol must be available in general purpose controller. And also helps in quick prototyping and deployment.

- Simple and user understandable addressing mechanism
  - Identifying the address of the node from its location is useful because it helps in keeping track of the network map of the complete system.
- The quality of service (QoS) guarantees
  - QoS helps to increase reliability of data transfer even in case of jitter, control latency and reduced data loss.
- Support wired as well as wireless connection
  - Network installation in existing buildings can be wired or wireless. To use existing infrastructure, protocol should be compatible with both.
- Availability of support by various programming languages
  - Different embedded devices are programmed in various programming language. Difference in the programming language should not effect communication between hardware.

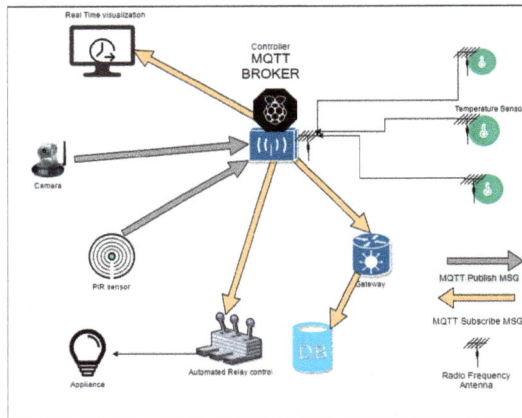

Fig. 3.5   MQTT communication in a system.

The Message Queuing Telemetry Transport (MQTT) protocol is found to be most suitable compared to others as per our design requirement. A comparison of the key features of various communication protocol can be found in Table 3.4. Sensors and Actuator nodes are efficiently addressed using this topic format. An address like "A/2/205/temp/ID" can be used to pinpoint a temperature sensor in Building A at second floor in room no. 205, type of sensor and its unique ID. Looking at the topic name, you can pinpoint the location of the sensors and the actuators. This protocol is suitable for group broadcasting; you can send a message to all second-floor

Table 3.4   Comparison of Various Communication Protocols.

Features	HTTP
Standardized by	IETF
Message Pattern	Request/Response
Messaging Type	Asynchronous & Synchronous
Reliability	Good
Addressing	URL
Constrained Devices	HTTP2
RESTful	Yes
Transport Layer Protocol	TCP/UDP
Security	TSL/SSL
Dynamic Discovery	No
Scalability	Complex

sensors to stop sending data when you do not want it. You can turn-ON all the lights of the first and third floors just by sending one command.

MQTT protocol has been implemented in many programming languages. This helps when different hardware programmed in different programming languages are able to communicate connecting to the same broker. All the popular IDE used for programming embedded devices have in-built packages to use.

MQTT in conjunction with node-red can give a complete visualization of the system. MQTT helps to generate a separate stream of data on demand, which can be used for data analysis purposes.

3.4.3   *Network Technology (WiFi)*

When it comes to installation of any building management system or just the sensors, the system must be economical, efficient enough to pay back in at least one and a half years. Installation of wireless sensor system will cost approximately 30% less compared to wired sensor installation which includes the cost of devices and labor charges. And maintenance will make it a lot worse. A wireless system will prove to be a lot better as the system is scaled up.

It provides a lot of technical advantages like mobility, distributed architecture, modularity, and re-usability. Let us see each one of them one by one. Things need not be fixed at one place. It is not necessary to hard-wire controller to the actuators. It can be kept anywhere as far as it is in the vicinity of the building wireless network. Wireless nature itself gives it a flavor of distributed architecture. Every module has a small controller which is sufficient enough to hold the state if there is any failure in other parts of the system. In such distributed system, we can eliminate complete system failure due to one point of failure. It is very easy to identify and replace malfunctioning modules. A failed module can be instantly replaced with a similar module with little reconfiguration over the visual interface.

The next question can be asked, why WiFi? Why not Bluetooth or Zigbee or any other wireless communication? WiFi has an advantage over other technologies as it is being used commonly by users at offices, homes, public spaces, etc. Other technologies need a specific type of upgrade to be made in infrastructure for proper functioning. There is one drawback that WiFi requires comparatively high power while communicating.

3.5   Analysis — Soft-Sensing and Energy Management in Buildings

Computational techniques can be used to i) detect and estimate occupancy, ii) find number of ON appliances in order to achieve the overall goal of reducing energy consumption. Occupancy sensing is important while providing thermal comfort in buildings because large amount of energy is wasted due to running of HVAC equipment during the period of non-occupancy.

3.5.1   *Soft Sensors: Non-Intrusive Monitoring (NIM)*

A soft sensor is a sensor that enables tracking of a facet through other existing sensors or data sources that were deployed for other reasons. The ability to sense a facet through these other means eliminates the need to deploy hard sensors for that facet, which saves on instrumentation costs. Soft sensors can be categorized into two types:

3.5.1.1   *Primary Soft Sensors*

Primary soft sensors reveal a facet without any additional inputs from any other source. Many primary soft sensors are deployed as part of a building's infrastructure or are part of the IT infrastructure within the building. For example, consider an office building that uses a swipe card system for entry and exit from the building. While the swipe card system is designed for access control — to only allow authorized users — it is also a soft sensor for occupancy since it reveals which users and how many users are present within the building. Similarly, since WiFi is ubiquitous within office buildings, the number of mobile smart phones connected to the various WiFi access points reveals the number of users within different areas of the building. Other examples of primary soft sensors include online calendars, room schedules, etc. We note that soft sensors are often approximations of the facets they reveal. In the above examples, visitors to a building may not have swipe cards or a user may choose to use cellular data on their phone without connecting it to WiFi.

3.5.1.2 *Secondary Soft Sensors*

Secondary soft sensors employ inference logic or algorithm which uses the data of one or more facets to obtain the value of another facet. For example, machine learning techniques can be used to develop an Inference Engine (*Inf*), which would take temperature and time as inputs and predict power consumption in providing thermal comfort. Similarly, for offices, *Inf* can take temperature and RFID data as inputs to predict power consumption due to HVAC equipments, lights and fans. Other Inference Engines like load dis-aggregation algorithms are examples of secondary soft sensors.

Examples of Soft Sensor Construction and Usage

• *Occupancy Detection and Counting:*
Occupancy Detection and Counting can be utilized as an intelligent aid in emergency evacuation, locating faulty appliances and many other application areas. Today, many sensors are available for occupancy detection/counting in different scenarios. But most of these sensors require user intervention, and are not scalable and are expensive. However, they can also damage the aesthetics of an existing building and hinder free user movement.

WiFi, calendar and messaging clients can be used as secondary soft sensors to help detect occupants [Ghai *et al.* (2012a); Kleiminger *et al.* (2013a)] in a commercial building and smartmeters can be used to detect occupants in a household. Let us look into how ARP scan,[1] screen lock and schedule can be used to detect occupants in the building. If the system is in screen saver mode, this status is communicated to a server.

The server does ARP scan of all machines in the department building every 15 minutes. The number of pings received from the ARP scan is the total number of active PCs in the building (x). The server also has the number of PCs in screen saver mode (y). The total number of occupants actively using the PC is ($x-y$). Other users of the building could be in classrooms or meetings which can be found using calendar data.

At a finer level — such as labs and offices, users register themselves

---

[1]ARP (Address Resolution Protocol) scan is a command line tool to count the number of active machines connected to the local network.

along with their desktop and/or laptop MAC ID and IP address as the current network policy dictates the use of static IPs. With this user-to-IP mapping, pinging the IP addresses gives the current status of the machine and the presence of a particular user in the lab or office.

• **Inference Engine Inf$_{\text{Reg}}$** for *Estimating Power Consumption based on Temperature and Occupancy Count*:

The power consumption in a building can be estimated using a regression model which takes temperature and occupancy count as inputs. In this case, the feature vector comprises of the *hour-of-the-day, mean of temperature and occupancy count in the past five minutes*. Figure 3.6 portrays the performance of the regression model from an experiment carried out in a smart classroom complex [Agarwal *et al.* (2016)] — it gives the actual and estimated consumption of a sample day in June. The model, which gives an estimation of power consumption every 5 minutes, has an accuracy of *91.7%*.

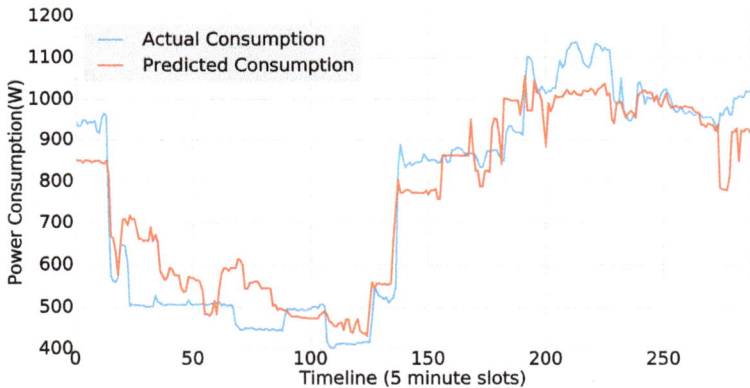

Fig. 3.6   Using Soft Sensor (Regression): Actual versus predicted power consumption. [Agarwal *et al.* (2016)]

• **Inference Engine Inf$_{\text{LD}}$** *for Load Dis-aggregation:*

This is an active research area, where the number of ON appliances and the category of appliances can be predicted by observing only the power consumption profile obtained from a central meter [Hart (1992a)] or sometimes context-aware information is used along with the meter data in order to produce better accuracy. It is evident that the number of ON appliances have a direct impact on energy consumption. In a smart building, on the other hand, appliances are often controlled for their optimum usage based on certain inferences drawn from some measured data. For example, lights, fans are switched ON when occupancy is detected and HVAC appliances are switched on based on certain temperature thresholds. Hence inference engines can be trained to derive many useful information like:

• **Inference Engine Inf$_{app}$** *for Identifying Number of* ON *Appliances:* This takes occupancy status and temperature as input and estimates the number of ON appliances for the corresponding period. The rationale behind this inference is that the temperature in an occupied room is indicative of the number of running HVAC equipments. For example, in a smart classroom complex (SCC), a state machine not only controls the appliances based on occupancy status, but also based on the temperature, to ensure users' thermal comfort.

• **Inference Engine Inf$_{frm}$** *for Instantaneous Power Consumption:* It takes the type and number of ON appliances as inputs and infers the instantaneous power consumption. In a smart classroom complex (SCC) case study (Sections 3.5.4.1 and 3.6), the inference engine is trained using a clamp-on meter for power consumption. Lights, Fans and IDUs[2] consume a constant power of 147, 47 and 40 Watts, respectively. The power consumption in a classroom C_{CR} is inferred by

$$C_{CR} = \left(\sum L_i * C_L\right) + \left(\sum F_i * C_F\right) + \left(\sum A_i * C_A\right)$$

where, C_{CR}, C_L, C_F, C_A denote instantaneous power consumption of classroom, light, fan and IDU, respectively.

L_i, F_i, A_i denote the number of lights, fans and IDUs respectively in classroom i.

3.5.2 *Achieving Desired Observability of Various Facets of a Building*

In order to achieve and enhance observability of various facets of a building, information about the structure of buildings can also be exploited, in addition to the semantics of various compositions of facets and sensors.

3.5.2.1 *Using Knowledge of the Structure of a Building*

Every building can be viewed as a tree, as shown in Figure 3.7, which represents an academic building. A node in this hierarchy is any part

---

[2]IDU here refers to the indoor unit of the HVAC, which is a constant power consuming device.

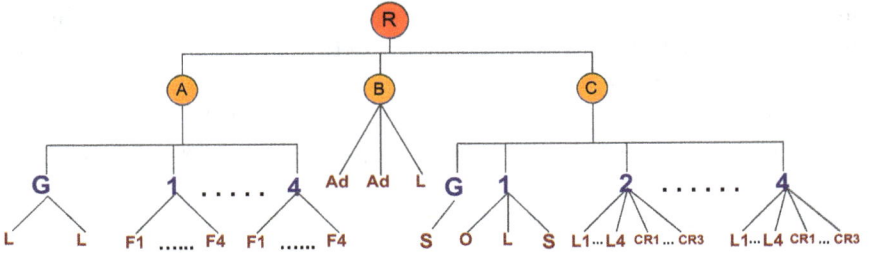

Fig. 3.7   Graph model of a building. [Agarwal *et al.* (2016)]

of the building which is required to be monitored, it could be a building (typically forming the root of the tree), a floor (marked as G,1,2,3,4), wing (marked as A,B,C) or a room. Leaves are typically the lowest level entities, like labs (marked as L), classrooms (marked as CR), faculty room (marked as F), office (marked as O), auditorium (marked as Ad) and server room (marked as S), which need to be monitored.

Rules for Inferring Observability of Facets from Other Facets and Sensors

- **Rule 1:** A facet F at the node N is observable if there is a sensor installed at N that observes F.
- **Rule 2:** If all the children of a node N are observable, then that node N is observable.

$$\forall C \in Child(N) Observable(C) \implies Observable(N)$$

  If all the children nodes are observable, then V_P, the value of the facet at the parent node P can be obtained as

$$V_P = f(V_{C_1}, V_{C_2}, ..., V_{C_n})$$

  where, f is an aggregation function that determines how V_P is computed from the value of N's children (V_{C_i}), see Section 3.5.2.2.
- **Rule 3:** If node M's parent and all siblings are observable, then M is observable. If N is the parent of M,

$$Observable(N) \wedge \forall C \in Siblings(M), Observable(C) \implies Observable(M)$$

  If the parent and the siblings of node M are observable, then V_M, the value of the node M is

$$V_M = g(V_P, V_{C_1}, V_{C_2}, \cdots, V_{M-1}, V_{M+1}, \cdots, V_{C_n})$$

where, g is a function which determines a facet of a child node based on the values of the siblings and the parent node. For some facets it may not be possible to uniquely determine the value of the child node from the parent and sibling nodes. For example, the value of the child node cannot be computed uniquely if the facet being observed is Occupancy status or Vacancy, which follow the logical OR and logical AND semantics, respectively.

3.5.2.2 *Using the Semantics of Composition of Facets*

According to Rule 2, the value of facet at a parent node is computed from the value of the facets at the children nodes, using the aggregation function f. Different types of aggregation and the facet, which can use them, are as follows.

- *Additive property* can be used to observe energy consumption by taking a sum of energy values from all the children nodes.
- *Average property* can be used for facets like temperature.
- *OR property* can be used by occupancy status.
- *AND property* can be used by vacancy status.
- *Min/Max property* may be used by temperature.

We can compute the value of a facet for a child based on the value for its siblings and the parent node, according to Rule 3. The following example computes the consumption of a child node M from parent and siblings' consumption.

$$V_M = V_P - [+(V_{C_1}, V_{C_2}, ..., V_{M-1}, V_{M+1}..., V_{C_n})]$$

Assuming that the building can be represented as a tree, observability of each node can be determined using the above rules.
- **Initialization:** From Rule 1 mentioned in Section 3.5.2.1, $Observable(N)$ is true if a sensor is attached to that node.
- **Bottom up Pass:** In this pass, by making use of Rule 2, node N's observability is affected by its children.

With this pass, non-leaf nodes that have all the children observable, will be marked observable.
- When the bottom up pass and observability rules are repeatedly executed, more nodes become observable, and is stopped until no further changes happen. At that point we have determined the observability of nodes in graph — given the initial configuration of (hard + soft) sensors.

We will discuss the applicability of rules stated above in Sections 3.5.4.1 and 3.6 to make the entire Smart Classroom Complex observable with respect to various facets.

3.5.2.3 *Exploiting the Facet-Sensor Relationship*

A facet-sensor relationship graph is used to represent the relationship between different facets and sensors in a building scenario.

- The nodes in this graph are of three types:
    - *Facet Node*: where facet value is observed
    - *Sensor Node*: which is represented by various sensors available. It can be a hard or a primary soft sensor node.
    - *Gate Node*: We have OR and AND gates.
- Edges represent the flow of information among the nodes. There are two types of edges, one which denotes the normal flow of information, another denotes flow of information through analysis, which are the secondary soft sensors.

Figure 3.8 shows the general facet-sensor relationship graph which consists of different observability facets, hard sensors, primary soft sensors, secondary soft sensors and the relationship between them. This graph is a generalized representation of the facet-sensor relationship in an academic building.

The facet-sensors relationship graph is used to establish a relationship between the facets and the sensors: if the value of facet F_1 can be used to infer the value of facet F_2, the cost of sensor deployment for observing F_2 is reduced, sometimes even eliminated.

In the figure, *Occupancy(#)* refers to the number of occupants, and *Occupancy (0/1)* refers to the occupancy status, i.e., occupied or unoccupied.

The value obtained from the primary soft sensors to observe occupancy count requires logical analysis to infer the exact value, hence the data is passed through the $LI_{oc\#}$. Similarly, value obtained from some hard sensors also require some kind of logical analysis to give the exact value of the facet. Therefore, the value obtained from Application Control Portal (ACP) is passed through LI_{app} and values obtained from either PIR or Camera are passed through LI_{ocS} to give the exact value of the facets they observe.

Consider the case of a laboratory, where the deployed/existing sensors are smartmeter, temperature sensor and ON PCs. In order to observe the

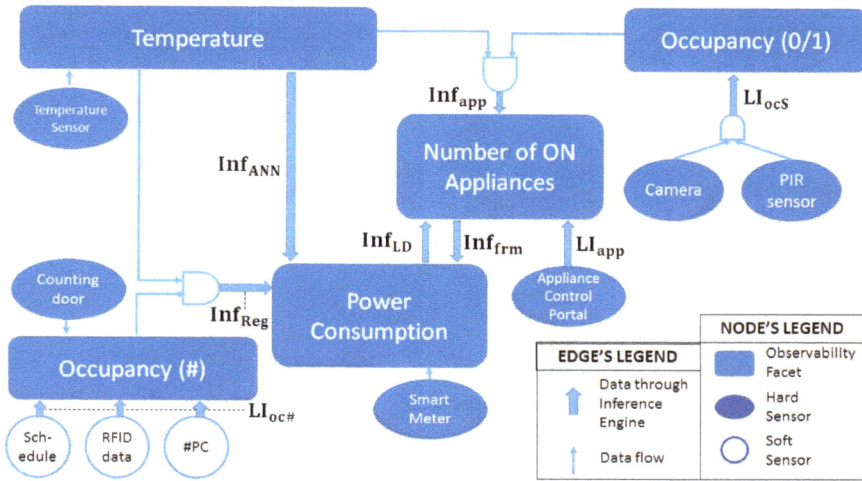

Fig. 3.8   A sample facet-sensor relation graph. [Agarwal *et al.* (2016)]

power consumption, a facet-sensor relationship graph (FSRG) can be constructed as follows.

- Consider power and temperature as facet nodes; smartmeter, ON PCs, and temperature sensor as sensor nodes.
- To observe power, use Inf_{ANN}, which uses ANN to infer power consumption based on zone temperature and acts as an input to the power facet node from temperature facet node. Similarly, information from ON PCs node and smartmeter nodes is provided as an input to the power facet node.

Thus, given the sensors available to observe a facet, its FSRG can be constructed by analyzing the relationship among the sensors and the facet(s).

In order to observe a facet F at a particular node N in a building, Algorithm 1 can be used. The algorithm exploits the facet-sensor relationship graph in exploring all the different sensors to obtain the value of a facet and selects the best sensor to observe the facet.

Consider the example of one wing, the *C-wing*, from the tree representation of the academic building in Figure 3.7. To observe power consumption and number of occupants at various levels in the C-wing, first check for the readily available information on different nodes (see Table 3.5). We

---

Algorithm 1 Algorithm to derive best value of a Facet F at a tree node N

1: **function** GETBESTFACETVALUE(F, N):

2:   $sensorList$ = list of all sensors that are adjacent to F in Figure 3.8 and are installed on N

3:   **for** each sensor s_i in the $sensorList$ **do**

4:     **if** s_i is a "secondary soft sensor" and all its inputs are not available **then**

5:       remove s_i from $sensorList$

6:     **end if**

7:   **end for**

8:   **if** $sensorList$ is empty **then**

9:     it is not possible to observe F

10:   **else**

11:     among the available sensors in the $sensorList$, choose the sensor with best accuracy. Let that sensor be s_i. $value = s_i.value$
    return $value$;

12:   **end if**

13: **end function**

Table 3.5   Existing information on various types of nodes.

Nodes	Available Information
Lab	ON PCs, Temperature sensor
Classroom	Occupancy Status, Temperature sensor
Server room	-
Office	Occupants' Schedule, RFID data, ON PC, Temperature

first need to run the algorithm on different types of nodes present in the C-wing. The C-wing has classrooms, labs, office and server rooms. In order to observe occupancy in a lab, Algorithm 1 consults Figure 3.8 and checks all the possible options to measure occupancy. The available data related to the lab are the number of ON PC's and temperature. Similarly, the algorithm can be applied for classrooms and office space. For server room, power consumption cannot be inferred because existing infrastructure does not support any instrumentation towards occupancy in the server room. Table 3.6 shows the various hard and soft sensors that can be chosen by Algorithm 1 to observe the various facets at different types of nodes.

Table 3.6 Sensors to observe facets.

Nodes	Power		Occupancy Count		Number of ON Appliances	
	Hard Sensor	*Soft Sensor*	*Hard Sensor*	*Soft Sensor*	*Hard Sensor*	*Soft Sensor*
Office	–	Inf_{Reg}	–	Schedule data	ACP	–
Lab	–	Inf_{ANN}	–	ON PC	ACP	–
Faculty room	–	Inf_{Reg}	–	Schedule data	–	ON PC, temperature
Server room	Smartmeter	–	–	–	–	–
Auditorium	Smartmeter	–	CD	–	ACP	–
Classroom	Smartmeter[3]	Inf_{frm}[3]	–	RFID data	–	Inf_{app}

3.5.3 Sensor Placement

A systematic approach towards placement of sensors is necessary so that the desired facets of a building can be made available using the i) existing hard sensors and ii) the soft-sensing technique. This approach can also help in reducing the number of sensors and yet offer full observability.

Refer to the C-wing *tree* of Figure 3.7. This wing consists of 31 nodes, of which 25 are leaf nodes, 5 are floor nodes and 1 is a wing node. Let us focus on tracking power consumption across C-wing. Consider two scenarios for placing sensors optimally across C-Wing i) when only hard sensors are present, and ii) when both hard and soft sensors are present.

3.5.3.1 *Using Only Hard Sensors*

There are two ways of placing the hard sensors to maximize the observability:

- *Placing hard sensors at all the nodes*: As the C-wing has 31 nodes, using this approach we need **31 hard sensors** in order to get complete observability of power consumption.
- *Placing sensors only at leaf nodes*: After placing sensors only at the leaf nodes of the tree, use the bottom up approach mentioned in Section 3.5.2.2 to obtain the value of non-leaf nodes i.e., the floor, wing and the building node(s). As C-wing has 25 leaf nodes, by following this approach we need only **25 hard sensors**. Thus, the number of hard sensors can be reduced, when placed at appropriate nodes, offering the desired observability.

---

[3]In the SCC, both the smartmeter and the Inf_{frm} are used to observe the power consumption.

3.5.3.2 *Using Both Hard and Soft Sensors*

The requirement of hard sensors can be reduced further by strategic use of soft sensors. The main disadvantage of using only hard sensors is that in order to obtain full observability, we need to procure hard sensors in large numbers and hence it will incur high initial investment as well as higher maintenance cost. Thus soft sensors, with a defined inaccuracy bound, can be used as an alternative and effective technique for cost reduction. To find the minimum number of hard sensors required to make the building observable, Algorithm 2 can be used.

Algorithm 2 Algorithm to find the minimum number of hard sensors to make the building observable for a facet F

1: **for** each node n in the leaf nodes of the tree **do**
2:     **if** GETBESTFACETVALUE(F, n) finds no sensor to observe F **then**
3:         We deploy a hard sensor to monitor F
4:     **end if**
5: **end for**
6: Make all the non-leaf nodes observable by using "Bottom up Approach", mentioned in Section 3.5.2.2

Note that Algorithm 2 internally uses Algorithm 1 for selecting the best set of sensors to be deployed in the building for achieving full observability. For applying Algorithm 2 on the entire building (Figure 3.7), the number of sensors required are summarized in Table 3.7. An important point of observation to note here is that in B-wing, there are two auditoriums, which have ad-hoc schedules and have no available information to observe the facets of interest. Hence, we need two smartmeters and two counting doors for B-wing. Whereas in A-wing, we have information available from labs and faculty rooms. Hence, we do not require any hard sensor in A-wing. The sensors selected to observe different facets on various nodes are listed in Table 3.6.

Table 3.7 shows the minimum number of hard sensors required to be placed in various wings and across the building to observe Power consumption and Occupancy. We observe that sensor placement strategy using Algorithm 2 requires minimum number of hard sensors to be deployed for achieving complete observability.

Table 3.7   Minimum number of hard sensors required to observe a facet in different wings.

Strategies	A-wing		B-wing		C-wing		Building	
	SM	CD	SM	CD	SM	CD	SM	CD
sensors on all nodes	24	24	4	4	31	31	60	60
sensors only on leaf nodes	18	18	3	3	25	25	46	46
using Algorithm 2	0	0	2	2	2	0	4	2

3.5.4   *A Case Study on Observability of Facets*

This case study involving a smart classroom complex (SCC) will demonstrate the applicability of the techniques of observing a facet of a building.

3.5.4.1   *The Smart Classroom Complex (SCC)*

The SCC is designed with the goal of ensuring a) zero occupancy implies zero energy consumption and b) minimal power consumption in an occupied room. If occupancy is detected, the lights are turned ON, and subsequently, fans and HVACs are turned ON based on the temperature. If room is detected to be vacant, all appliances are switched OFF. Refer to Figure 3.9 for the schematic of the Smart Classroom Complex (SCC) in the form of a tree diagram. The root node 0 represents the SCC. Nodes 1 and 5 denote small classrooms; nodes 2 and 6 denote big classrooms. Nodes 3 and 4 denote the zones of big room 2; similar relationship exists between nodes 7, 8 and 9. The HVAC system consists of:

- IDU: indoor unit of HVAC, located inside the classrooms.
- ODU: outdoor unit of HVAC, located outside the classrooms and controls multiple IDU's in the SCC. SCC's schematic has a node 9 (*ODU* node) for denoting the ODU present in the SCC. The set temperature

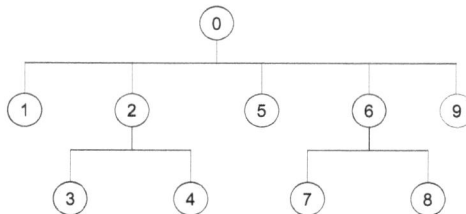

Fig. 3.9   Tree representation of SCC.

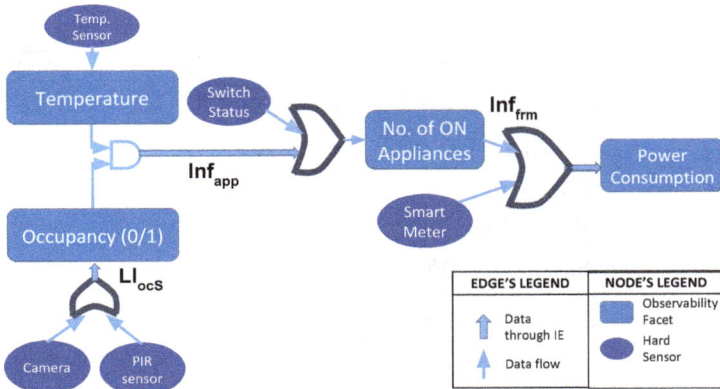

Fig. 3.10   Facet sensor relationship graph of SCC. [Agarwal *et al.* (2016)]

in the IDU's and the number of operational IDU's at any given time determine the cooling needed and hence the power consumption of the ODU.

3.5.5   *Observing Facets of Interest*

In order to achieve the goals of SCC, the facets to be observed are the following.

- **Power Consumption**: One of the design goals of the SCC is to monitor and minimize power consumption. In addition, monitoring power consumption can help us in identifying faulty appliances.

  Note that installing hard sensors on all the leaf nodes would require a total of seven smartmeters to achieve the goal.

  To minimize the requisite number of smart meters, Algorithm 2 is applied. It chooses Inf_{frm} for all the leaf nodes except the node of type ODU. Hence, only one smartmeter is required to make all the leaf nodes of SCC observable. Once all the leaf nodes are observable, the bottom-up approach using the facet semantic **SUM** is applied making all the non-leaf nodes observable. Thereby, all the nodes of SCC become observable with the help of a single smart meter.

  The case study shows that power consumption in each room can be observed using only one smart meter (Figure 3.11), if Algorithm 2 is used. This reduces the cost drastically.

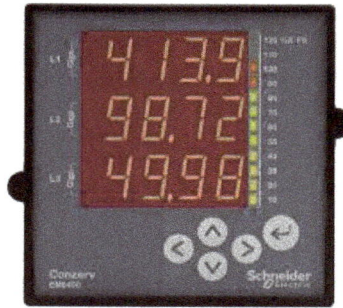

Fig. 3.11   Smart power meter (Image Courtesy: schneider-electric.co.in).

- **Occupancy Status**: This helps in decision making for controlling the classroom appliances. Since occupancy status information satisfies the requirement, we chose to observe occupancy status rather than occupancy count (though the former can be inferred from the latter) in order to reduce the cost of deploying sensors.

  Since all the classroom nodes are observable using PIR and Camera, occupancy status of the SCC can be obtained by taking an *"OR"* of the status values from leaf nodes upward to the root.

  The vacancy status is simply the *"AND"* of the vacancy status values of all the leaf nodes (NAND of the occupancy status).

- **Temperature**: Maintaining user comfort is also one of the critical goals of SCC, failure in which erodes the benefits from the smartness of the system. Maintaining user comfort is a critical element in realizing the benefits derived from the smartness of the SCC.

  Fans and IDUs are controlled based on real-time temperature values to make the occupants feel comfortable while minimizing the power consumption when there is occupancy. Temperature sensors have to be deployed in all zones that need to be monitored for thermal comfort. To find the average temperature of a big classroom, *Weighted Average* of the temperature values of all the zones of the big classroom can be used.

- **Number of ON Appliances**: SCC provides us with zonal occupancy data and hence, it is possible to turn ON only those appliances which are in the occupied zones. Tracking the number of ON appliances confirms if the actuation is properly executed and also check if the energy consumption is only due to the desired appliances. Here, appliances comprise of lights and fans. Inf_{app} is used to observe the number of ON appliances. Also, there are hard sensors that can give real-time status (ON/OFF) of the appliances. For observing the number of ON appliances at root node, we can SUM the values from the child nodes.

3.5.5.1 *Techniques used in Observing Facets of SCC*

The facet sensor relationship graph with respect to SCC is already shown in Figure 3.10. In the figure, the edge labeled LI_{sum} denotes that the values passing through it get summed. The facet sensor relationship graph shows the relationship between different facets of interest and (hard and soft) sensors that can be used to observe these facets. By applying Algorithm 2 on this graph, we can get the final list of sensors to observe different facets in SCC, as shown in Table 3.8. Let us now look into the details of the techniques used.

Inputs

- $\{N\} = \{0, 1, ...9\}$ — nodes in the tree.
- $\{F\} = \{Power, App, Occ, Temp\}$ — facets to be observed are power consumption, number of ON appliances, occupancy status and temperature.
- $\{t^{Power}\} = \{sm\}$ (smartmeter), $\{t^{App}\} = \{sws\}$ (switch status), $\{t^{Occ}\} = \{PIR, camera\}$, and $\{t^{Temp}\} = \{Tsensor\}$ (temperature sensor).
- *facet requirements*: all the facets should be observed on all nodes, except node 9.
  $\forall n \in \{\{N\}\backslash\{9\}\} \ \forall f \in \{F\} \qquad need\_observe(n, f) = 1$
  On node 9, only power consumption should be observed.
  $\forall n \in \{9\} \ \forall f \in \{Power\} \qquad need\_observe(n, f) = 1$
  Number of ON appliances, occupancy status and temperature should not be observed on node 9.

Table 3.8   Sensors required for observing facets of interest in smart classroom complex.

Facet	Hard Sensors	Soft Sensors
Power	SmartMeter	Inf_{frm}
Occupancy Status	PIR, Camera	-
Number of ON appliances	-	Inf_{app}
Temperature	Temperature Sensor	-

$$\forall n \in \{9\} \ \forall f \in \{\{F\}\backslash\{Power\}\} \qquad need\_observe(n, f) = 0$$

- *FSRG*: Figure 3.10
- *Tree representation*: Figure 3.9

Facet Requirements:

The final observability status of all the facets on all the nodes ($observable(n, f)$) should be equal to the facet requirements specified at each node.

$$\forall n \in \{N\} \quad \forall f \in \{F\} \qquad observable(n, f) = need\_observe(n, f) \quad (3.1)$$

Using FSRG to observe a facet:

Given the FSRG as in Figure 3.10, let us assume that $f1$ is a facet that can be observed using sensor $s1$ or inferred from facets $f2$ and $f3$. Thus $observable(n,f1)$, the *observability* of $f1$ at a node n can be stated as

$$
\begin{aligned}
observable(n, f1) = &\ used(n, f1, t_{s1}^{f1}) \ OR \\
&\ (observable(n, f2) \ AND \ observable(n, f2)) \quad (3.2)
\end{aligned}
$$

where, $used(n, f1, t_{s1}^{f1})$ represents the observability of facet $F1$ of node n using sensor $s1$ placed at n.

Observability Rules to observe a facet: a facet is appendable at a node if a sensor that observes the facet is installed on the node, or all its children nodes observe the facet (Section 3.5.2.1). Let p be a node and $\{c_1^p, ...c_c^p\}$ be the set of children nodes of node p (set of children nodes is empty if the node p is a leaf node). *A facet f at the node p can be observed by either placing a sensor $s1$ at the node ($used(p, f, t_{s1}^f)$) or aggregating the values from all the children nodes (AND of observable of all children nodes).* Hence we use *Rule 1* and *Rule 2* to observe the facet at the node.

$$
\begin{aligned}
observable(p, f) = &\ used(p, f, t_{s1}^f) \ OR \\
&\ (observable(c_1^p, f) \ AND \ observable(c_2^p, f) \quad (3.3) \\
&\ \cdots \ AND \ observable(c_c^p, f))
\end{aligned}
$$

Application of Observability Constraints and Rules in the Case Study

(1) *Satisfying Facet Requirements:* all facets are required to be observed on all the nodes except node 9, where only power consumption should be observed. This requirement is specified as part of the input. The constraint is identical to constraint 3.1.

(2) *Using FSRG to observe facets:* We will now study how constraint 3.2 is applied in the formulation of constraints for applying in SCC. The reader is referred to Figure 3.10 to determine how each facet can be observed in different possible ways.

(a) *Power consumption:* power consumption is observable if a smart-meter sm is placed on node n ($used(n, Power, t_{sm}^{Power})$), or the facet *number of ON appliances* is observable on node n ($observable(n, App)$). Formally, the constraint is represented as:
$\forall n \in \{N\}$:

$$observable(n, Power)$$
$$= used(n, Power, t_{sm}^{Power}) \ OR \ observable(n, App) \qquad (3.4)$$

Writing OR in the constraint equation format is presented below:
Let $\forall n \in \{N\}$:

$$observable(n, Power) <= used(n, Power, t_{sm}^{Power}) + observable(n, App)$$

$$observable(n, Power) >= used(n, Power, t_{sm}^{Power})$$

$$observable(n, Power) >= observable(n, App)$$

(b) *Number of ON Appliances (App):* it is observable at node n if a switch status sensor sws is placed on n ($used(n, App, t_{sws}^{App})$), OR occupancy status and temperature facets are *both* observable at node n ($observable(n, Occ) \ AND \ observable(n, Temp)$). The constraint is formulated as:
$\forall n \in \{N\}$:

$$observable(n, App) = used(n, App, t_{sws}^{App}) \ OR$$
$$(observable(n, Occ) \ AND \ observable(n, Temp)) \qquad (3.5)$$

Function AND in the constraint equation format is presented below:
Let $\forall n \in \{N\}$:

$$x1 = used(n, App, t_{sws}^{App}),$$
$$x2 = observable(n, Occ), \ x3 = observable(n, Temp)$$

Let us take a variable $x4$ to present the AND construct (in constraint equation format), which infers the number of ON appliances from observable occupancy (Occ) and temperature (Temp). The AND between $x2$ and $x3$ can be denoted by the following constraints, which ensures that $x4$ can be 1 only if both $x2$ and $x3$ are 1.

$$x4 \geq x2 + x3 - 1, \quad x4 \leq x2, \quad x4 \leq x3$$

The *OR* between $x1$ and $x4$ is computed to obtain the final observability status of number of ON appliances at node n.

$$observable(n, App) <= x1 + x4$$

$$observable(n, App) >= x1, observable(n, App) >= x4$$

(c) *Occupancy Status and Temperature:* occupancy is observed using a PIR sensor or a camera. Temperature is observed using a temperature sensor *Tsensor*.

$$\forall n \in \{N\} \quad \forall f \in \{Occ, Temp\} \quad observable(n, f) = \sum_{t \in \{f\}} used(n, f, t)$$

(3.6)

3.6 *Respond (Timely) using Hybrid Sensing: a Case Study*

Let us take the Smart Classroom Complex (SCC), already discussed in Section 3.5.4.1, for the case study. This is a system that aims at providing desired habitability in terms of thermal comfort and illumination while minimizing power consumption.

The devices/appliances such as lights, fans and HVACs in SCC are supposed to be switched-ON for 5 hours and 30 minutes i.e. 8:30 AM to 12:30 PM and 2:00 PM to 3:30 PM. Students or staff of the building are expected to turn the appliances OFF during the one and half hour break. But, in reality, it has been observed that appliances were never turned off for the entire seven hours which resulted in energy wastage. Therefore, it became necessary to collect user activity pattern by sensing their movement and accordingly control the devices (HVACs, Fans, Lights), which can be achieved using hard and soft sensing mechanism. This is achieved by observing other facets like occupancy status and temperature. On detection of occupancy, lights and fans are turned ON, and subsequently, fans and HVACs are turned on based on temperature. If the room is detected to be vacant, all the appliances are switched OFF. This is explained using the State Diagram in Figure 3.12. As it can be observed from Figure 3.12, the possible states of the classroom are

- **Empty**: When the room is unoccupied, i.e., when human not detected (HND),
- **OCC+L**: When occupancy is detected and only lights are switched ON,

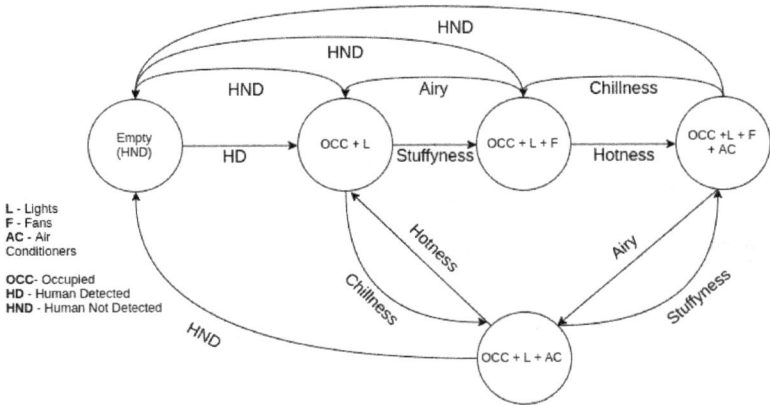

Fig. 3.12   SCC state diagram.

- **OCC+L+F**: When both lights and fans are ON in the occupied room,
- **OCC+L+F+AC**: When AC is also required to maintain thermal comfort of the occupants in the room.

Implementation of the above state machine ensures saving of energy without compromising the illumination and thermal comfort requirements of the classroom occupants.

3.7   Summary and Takeaways

In this chapter, we discussed how energy management technologies can be embedded within commercial and residential buildings. We presented building management systems and networking protocols used for energy management in commercial buildings. We also presented IoT devices, sensors and protocols that are commonly used in residential buildings for energy management and automation. We then described a principled approach for determining what sensors to deploy as well as how many and where to deploy them to achieve various goals. We also discussed how soft sensors can be exploited to reduce the cost of deploying hard sensors in a building. We provided a characterization of different types of building loads and how to model these loads and ended with algorithms and case studies for optimizing the cost of provisioning sensors within buildings.

Building energy management remains an active area within the research community and industry. There is ongoing work on new protocols that enable compatibility with mobile devices as well as enable interoperability with Internet protocols. New low-cost sensors as well as new inference methods for soft sensing remain an active area. Finally, algorithms that fuse data from various sensors to infer the state of building and take timely actions also provide new opportunities for further improvements.

The next chapter is dedicated to a holistic approach for achieving thermal comfort in buildings following the SMART cycle of Energy Management.

Chapter 4

A Systematic Approach to Thermal Comfort

4.1 Introduction

People spend majority of their time in buildings and 40% to 70% [DoEE (2017)] of the energy consumed is spent on providing thermal comfort to the occupants. Thermal comfort in buildings is a necessity in order to i) meet growing demand for certain quality of habitability in workplaces as well as in homes and ii) improve the working efficiency of the individuals.

Rapidly growing energy demand in buildings is a major concern as it constitutes about 40% of total energy demand. Our dependency on non-renewable and polluting energy sources (i.e., fossil fuels) to meet the gross as well as peak demand have raised concerns over poor quality of service (occurrences of black-outs, brown-outs and load shedding), depletion of energy resources and its impact on the environment. Therefore, we need technological solutions, which will help not only in reducing consumption, but also in preventing wastage of energy while providing thermal comfort in buildings.

Any endeavour to optimize energy usage while providing thermal comfort calls for provisioning of smart solutions in today's buildings. The widely acknowledged solution involves transitioning existing buildings to "smart buildings" and making new buildings *born smart*. The process of making a building smart requires the replacement of aging thermal conditioning resources, widespread and judicious deployment of sensors and actuators, integration of renewable energy sources and automating building management using distributed Information and Communication Technology (ICT) systems. The Building Management System (BMS) should automate the process of maintaining thermal comfort involving heating, ventilation, air conditioning. Additionally, a BMS should also automate the process of managing lighting, security, and other systems like alerts and evacuation support system in case of emergencies like fire, earthquake, terrorist attacks, etc.

We begin this chapter with an introduction to the physical parameters that constitute thermal comfort, the techniques and the resources used to control these parameters in order to provide the desired thermal comfort to humans in a building.

At the end of this chapter, we also introduce the challenges involved in providing thermal comfort while simultaneously reducing consumption and preventing wastage.

4.1.1 *Thermal Conditioning for Individual Comfort*

At the outset, we outline the fundamental concepts and facts that help understand what constitutes a thermally comfortable environment from an individual's point of view.

4.1.1.1 *What is Thermal Comfort?*

ASHRAE standard 55 [ASHRAE (1992, 2001)] states that thermal comfort is a *state of mind*.

> **Thermal Comfort**
>
> "Thermal comfort is that condition of mind that expresses satisfaction with the thermal environment"

This is an open definition left to human cognition, but it clearly indicates that a consumer is not expected to express thermal comfort in terms of physical parameters.

However, in order to provide thermal comfort, it needs to be expressed in terms of controllable physical parameters. So, what are those parameters that constitute thermal environment in a given space? While this question appears deceptively simple, the answer to this has practical and far reaching implications on the design of systems that provide thermal comfort in building spaces.

Fanger [Fanger (1970)] identified four environmental parameters, along with one physiological parameter (the activity level) and one thermal resistance parameter (the clothing insulation), which constitute the thermal environment.

> **The Constituent Parameters of Thermal Comfort**
>
> (1) air temperature,
> (2) mean radiant temperature,
> (3) relative air velocity,
> (4) vapor pressure (a measure of humidity level) in the ambient air,
> (5) activity level i.e., internal heat production in the body, and
> (6) thermal resistance of clothing.

Human metabolic thermo-regulatory system maintains a reasonably constant deep body temperature and therefore, the thermal comfort of a person is related to the balance between heat dissipated or lost to the environment and the heat produced in the body. In [Fanger (1973)], it is also identified that

- People are not alike thermally, or otherwise. People, due to their biological variance like metabolism rate and body types can experience different levels of thermal comfort when exposed to the same climatic condition in a space. Therefore, it is normally not possible to satisfy everyone at the same time.

- Human beings cannot become adapted to prefer warmer or colder environments due to their exposure to colder or hotter surroundings, like in cold or tropical countries. Therefore, the same comfort conditions can be applied throughout the world.

Note that this finding was under contrived setting of the climate chamber and therefore it is debated by many researchers and practitioners who established that non-thermal factors like climate setting, social conditioning, economic considerations and other contextual factors do affect the thermal perception of the individuals. More on this topic can be found in Chapter 5.

4.1.2   *Thermal Conditioning for Server "Comfort"*

Next to human beings, it is the modern day servers, which are a significant number of consumers of electricity towards achieving its desired "thermal comfort", if we can call it so. Koomey's landmark analysis in a 2007 estimation study showed that 0.8% of total electricity was consumed by the servers worldwide. This has grown by many times today. Examples of machines, whose "thermal comfort" is an essential requirement, are traditional lab servers, ATMs, Bitcoin mining farms, etc.

However, in case of servers, the thermal comfort level is not a "state of mind", but a defined range of temperature and relative humidity. The range of temperature in server rooms and data centres are usually maintained from $17°$ to $27°C$ and the range of relative humidity (RH) is from 40 to 60%.

Therefore, there is room for playing with server room temperature in dealing with demand-response (D-R) mismatch.[1]

The techniques for D-R control that manipulate thermal comfort band for human occupants are applicable in the maintenance of thermal comfort in server rooms as well.

4.1.3 *Thermal Conditioning Resources*

Temperature is the major parameter, which is required to be maintained within desired limits in order to provide thermal comfort. ASHARE standard 55-2013 states that temperature for providing thermal comfort to human occupants could range from 67 to $82°F$ ($19.4°C$ to $27.7°C$) which depends on the relative humidity and other factors like season, clothing level, activity levels, etc. Therefore, both the temperature and the humidity of a closed space is controlled to provide thermal comfort to the occupants.

Air-conditioners offer cooling and reduce humidity levels at the same time. But have a higher run-time cost compared to humidity control appliances. Humidity control appliances are useful only in regions with extreme humidity levels and it is to be noted that their capacity to reduce thermal discomfort is limited.

A large fraction of the electricity bill in a building is due to the high amount of energy consumed by the conditioning resources like air-conditioners and heaters.

In order to reduce consumption while providing thermal comfort in a closed space, the following measures can be taken.

• Maximizing ventilation using operable windows, small table fans or

---

[1]It is important in power systems to maintain the balance between power generation and consumption. In other words, the amount of power generated must be consumed at any point of time. Following any mismatch, corrective action is taken by varying the power generation, or the consumption in order to restore the balance. This is termed as demand-response (D-R) control.

ceiling fans in moderately hot climatic areas, without using any thermal conditioning appliances.

- Judicious use of conditioning resources — switching off AC(s) when not required, running fewer AC units under low occupancy in a space equipped with more than one AC.
- Active consideration of climatic conditions, incident sunlight, shadowing effects and wind patterns in the design of new buildings.
- Improving insulation in both old and new buildings.

While proper ventilation by operable windows and the use of fans enhance thermal comfort (in hot climate) to some extent due to increased air velocity, air-conditioners and heaters remain the most effective means of achieving thermal comfort. Dehumidifiers and air-coolers are also used, but are not very effective due to their dependency on humidity levels and ambient temperatures. For example, air-coolers are found to be effective in areas with arid weather.

Further, dependency on voluntary human interventions, like switching off ACs when not required, may not be always practical to achieve the goal of reducing consumption. Also, participation of consumers in demand response (D-R) management cannot be guaranteed in spite of financial incentives. Therefore, it calls for measures like i) automated interventions during periods of non-occupancy and higher peak demand, ii) providing online information to the user on the financial benefits of reducing consumption during certain periods of the day just by going for a slightly inferior thermal comfort level, say $\pm 1°C$ from the desired temperature.

Since automated intervention demands the understanding of the equipment to be controlled, we introduce various types of air-conditioning resources, humidifiers, de-humidifiers and their fundamental operating principles. This will provide sufficient background for better understanding the topics discussed in this book.

4.1.3.1 *Air-Conditioners (AC)*

The thermodynamic principles of air-conditioning and refrigeration is the same for all types of air-conditioners, usually available in installations viz., i) Stand alone Window AC or split AC, ii) Variable Refrigerant Flow (VRF) AC and iii) Centralized air-conditioning system that use chiller plants.

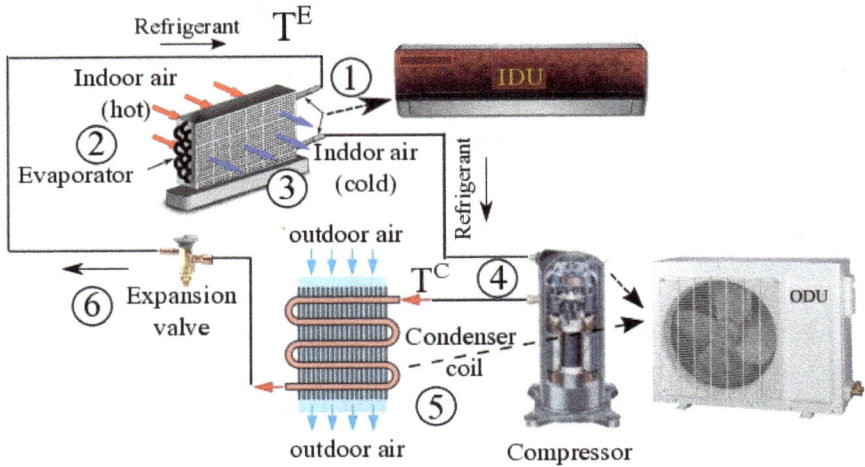

Fig. 4.1   Working of a typical air-conditioner.

Working of an AC

Let us look into the fundamental thermodynamic processes involved in a typical air-conditioner. As shown in Figure 4.1, the following steps are involved when an AC cools the air in a closed space.

(1) The cold *refrigerant* passing through the internal cooling coil (also called *evaporator*) inside the in-door unit (IDU) absorbs heat from the room air. The heat transfer between air and the refrigerant is assisted by a fan that draws air from the closed space and forces it to flow over the cooling coil. Thus, the air gets cooled and brings down the temperature of the closed space. By absorbing heat, the refrigerant boils up to vapor, which goes to the suction of the compressor. (Steps $1, 2$ and 3)

(2) Compressor inside the out-door unit (ODU) compresses the vapor and passes it to the inlet of external condenser coil. (Step 4)

(3) The vapor, while passing through the heat-exchanger (also known as *condenser*) gets cooled by forced (by fan) air in the outdoor unit (ODU) of the AC and gets condensed into high temperature, high pressure liquid which goes to the expansion valve. (Step 5)

(4) *Adiabatic expansion* takes place at the *expansion* or *throttle* valve. High temperature high pressure refrigerant (liquid) flashes to a mixture of liquid and vapor. The temperature of the refrigerant drops down a very

low value due to adiabatic expansion (Step 6) and the cold refrigerant enters the evaporator (Step 1). The process repeats.

A brief introduction to the fundamental thermodynamic principles behind the cooling provided in an AC is presented in Appendix C.1.

Window AC

A stand alone windows AC is a compact single unit that houses all its components viz., i) the compressor, its condenser coil and the external fan, and ii) evaporator coil along with the internal fan.

A Windows AC is controlled using a thermostat based ON-OFF or bang-bang controller. This type of an AC maintains the temperature of a closed space within a band, usually within $\pm 1°C$ of the set temperature. The compressor is cut-off when the room temperature goes below the set value and when it goes above, the compressor is restarted.

Nowadays, people prefer split-ACs for their low noise. In a split-AC, the condenser unit along with its compressor is installed outside the room and the evaporator unit is installed inside. By splitting the condenser unit from the rest of the components and keeping it outside, the noise produced by the compressor is eliminated from the space inside. Today, split-ACs for home applications can have an ON-OFF controller or a variable refrigerant flow (VRF) controller, popularly known as the inverter AC. However, inverter ACs are much more expensive compared to conventional ON-OFF ACs.

VRF AC

Variable Refrigerant Flow (VRF) ACs, which are gaining more popularity among domestic consumers, are more energy efficient than the ON-OFF controlled window ACs, but demand higher initial investment. VRF is a preferred technology for industrial or institutional applications that require centralized cooling of multiple rooms/spaces and when the installation of a chiller plant is not a viable option.

When a VRF AC is switched ON, it runs the compressor at its full speed to cool the space as fast as possible. Once the set temperature is achieved, the compressor is not switched off as in the case of a conventional ON-OFF AC, but allowed to run at a lower speed. This is because, once the air inside the closed space gets cooled to the desired value, a lower rate of cooling is required. This lower rate is decided by the requirement of cooling that would be enough to compensate for the heat produced (by humans and the heat producing equipment like computers, etc.) and the loss due to leakage.

But, the main advantage of a VRF system is that it can cater to a wide variation in the heat loads due to non-occupancy in some of the spaces at different points of time.

The controller of a VRF system uses variable speed drive to control the speed of the compressor and it receives the space temperature as the feedback signal. A VRF system can have one or more outdoor units (ODU), each one of which houses its compressor and the condenser. The cooled refrigerant flows through the various indoor units (IDU) and the air inside their respective spaces gets cooled when it comes in contact with the evaporator coil in the IDU, which is assisted by a fan. The flow of refrigerant through a particular IDU is controlled using its control valve and it is governed by its set temperature and the heat produced inside the space.

When a VRF system has more than one ODU, usually one ODU is not equipped with a variable speed drive, which is known as the base ODU. This is done based on the assumption that the VRF system will see a minimum heat load whenever it is switched on. Therefore, under very low load, a VRF system can over-cool because it has been observed that i) the base ODU runs only at its full speed and ii) it does not switch off its compressor till it runs for a pre-set duration, even if the space temperature reaches the set value.

Centralized Air-Conditioning with Chiller Plants

Centralized air conditioning system, as the name suggests, is used to cool a number of closed spaces using a centralized plant. Centralized air-conditioning system for large number of spaces mostly use chiller plants. VRF ACs are also used for centralized cooling, but limited to smaller number of spaces/rooms, because of its higher cost.

A chiller plant consists of at least two heat removal circuit — i) The chiller circuit and ii) the chilled water circuit. The condenser in the chiller circuit is often water-cooled, but it can be air-cooled also. A chiller plant with water cooled condenser involves a third circuit with a cooling tower and a pump, which supplies water to the condenser. A typical chiller plant schematic is shown in Figure 4.2.

The *chiller circuit* contains refrigerant and like any other air-conditioning system, it has its own compressor, the condenser and the

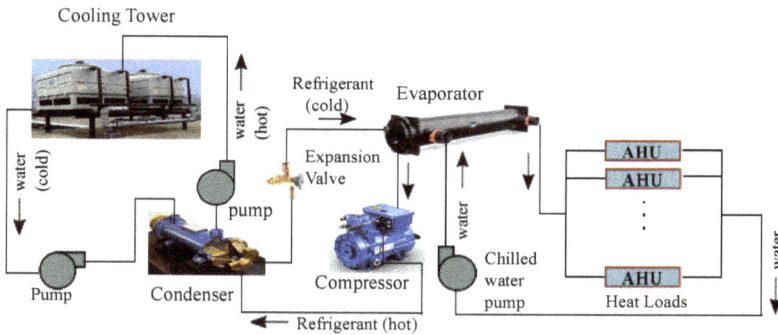

Fig. 4.2   Chiller plant.

evaporator. It is in the evaporator, the refrigerant absorbs heat from the water in the *chilled water circuit* and makes it chilled.

The chilled water is circulated through a number of air-handling units (AHU) running chilled water pipes in various places, using pump(s). The pipes carrying chilled water branches off into pipes with smaller diameter before entering into the AHUs. The space air gets cooled in the AHU by circulating air over the cooling coils, using fans. The chilled water absorbs heat in the AHU and the hot water is circulated back to the evaporator with the help of a pump, where it again gets chilled.

In the case of a water-cooled condenser in the chiller circuit, an additional heat removal circuit with a cooling tower is deployed. The heat absorbed by water at the condenser is removed at the cooling tower, with the help of a pump.

The chilled water circuit contains a liquid cheaper than a refrigerant, mostly water or glycol-water mix, as the heat transfer media.

The refrigerant in the chiller circuit transforms into vapor by absorbing heat in the evaporator, which is then compressed by the compressor and its temperature rises further. The hot and high pressure refrigerant vapor passes through the condenser, where it is cooled to its liquid phase either by forced air or by water. The hot and high pressure liquid then passes through an expansion valve. Here, due to adiabatic expansion, it gets transformed into cold mixture of liquid and vapor, which enters the evaporator, where the cold refrigerant again absorbs heat from the water in the chiller water circuit.

4.1.3.2 De-Humidification and Humidification

We have discussed previously that thermal comfort level also depends on humidity level. Though the ASHRAE standard 55-2013 standard does not specify a lower limit on humidity levels, it mentions that non-thermal comfort factors such as skin drying, irritation of mucus membranes, dry eyes, and static electricity may limit the acceptable lower thresholds of humidity. However, the recommended range of humidity level for human beings in a closed space is between 40% to 65%.

Humidification Techniques

Two types of humidifiers are available commercially, viz., *Evaporative* and *steam-based.*

Evaporative or *cool mist* humidifiers use a fan or a blower to evaporate water from its storage tank and throw it out into the room air through porous filters. The other type of evaporative humidifier uses rotating discs to evaporate and blow moisture into the air. It is interesting to note that *air-coolers* or *desert-coolers* popular in tropical countries are, in effect, evaporative humidifiers. They are also effective coolers only in dry climate areas. This is because evaporation of water draws its required latent heat from the room air and thereby cools it and the increase in humidity causes no discomfort in dry climatic condition.

Steam-based or *warm mist* humidifiers, on the other hand, release steam into the air by boiling water. Steam-based humidifiers are useful in cold climate where heat coming out from the boiling water system does not cause discomfort.

However, note that higher humidity level can cause the growth of microbial or mold causing damage to the building interior/infrastructure and more so to the building envelope made of natural materials. Due to this damaging effect of humidification, ASHARE standard 62.1-2016 recommends limiting humidification below 65%.

De-humidification Techniques

The common dehumidifiers used for thermal comfort in buildings are of the following types
- Refrigerative,
- Desiccant, and
- Thermo-electric.

Refrigerative Dehumidifier:

Refrigerative or *compressor type* is the most common type of dehumidifier. The working principle is the same as in a refrigerator or an air-conditioner where a compressed refrigerant undergoes adiabatic expansion and cools metallic evaporator coils. In a dehumidifier, the humid air passes over this cold evaporator coils by forced circulation using fans or blowers. Thus, moisture in the air condenses when it comes in contact with the cold coils and the air that exhausts out of the dehumidifier becomes drier than when it entered. The condensed water is collected and drained away.

Desiccant Dehumidifier:

Desiccant is a chemical, which absorbs moisture. Instead of using condensation technique to remove humidity from air, a desiccant dehumidifier uses a desiccant e.g., silica-gel. Air flows through a rotor containing chemical desiccant and the dry air flows back to the enclosed space or room. However, a desiccant can only be reused if the absorbed moisture is removed by heating and the generated water vapour is ducted outside the building, which incurs additional running cost.

Thermo-electric Dehumidifier:

Thermo-electric dehumidifiers makes use of the *Peltier effect* to facilitate dehumidification. Peltier effect[2] is a physical phenomenon named after the French physicist Jean Charles Athanase Peltier who observed that when a voltage is applied across the two dissimilar electric conductors connected at their ends, it created a temperature difference between the two junctions of these conductors. This difference in temperature created in a Peltier module, which has one cold side and one hot side, is utilised for dehumidification.

In a thermo-electric dehumidifier, a Peltier module is placed between two heat sinks in direct contact with the two junctions (hot and cold) of the module. Hot humid air is sucked from the hot side heat sink using a fan and forced through the cold side heat sink where the moisture present in the air gets condensed. Thus the air that comes out from the dehumidifier through the cold side heat sink is drier than what entered the dehumidifier through the hot side heat sink. The mechanical design (smooth finish angled fins) of the cold side heat sink facilitates the condensate to drip down easily and get drained away.

---

[2]This is opposite to the *Seebeck effect* named after German physicist Thomas Johann Seebeck, which is a phenomenon that produces voltage when there is a difference in temperature between the junctions of two dissimilar electric conductors.

The *advantages* of thermo-electric dehumidifiers are that they are quiet (as there is no noisy compressor) and reliable. However, they are less efficient than refrigerative dehumidifiers.

Air-Conditioning and Dehumidification

When an air-conditioner (AC) works in cooling mode, it also doubles as a refrigerative dehumidifier. The air gets cooled by coming in contact with the cooling (evaporator) coil of the AC and at the same time moisture in the air also condenses, which is drained outside the building.

In this context, it is worth noting that a Peltier cooler works on the same principle as Peltier (thermo-electric) dehumidifier.

Thermal comfort in buildings is a necessity in order to i) meet the growing demand for certain quality of habitability in workplace as well as in homes and ii) improve the working efficiency of the individuals.

4.2   Challenges in Providing Thermal Comfort in a Building

The provisioning for thermal conditioning in a building is usually made considering the worst case scenario of maximum thermal load on the hottest (or coldest) day of the year. This is largely done going by the conventional design approach using empirical formulae. It often leads to over-provisioning or under-provisioning. Further, buildings are hardly designed considering thermal conditioning as a requirement. Therefore, existing provisions may not always guarantee the desired thermal comfort with minimum energy consumption. Further, considerable amount of energy gets wasted as it is not guaranteed that consumers will always be energy conscious and use the thermal conditioning resources judiciously. Hence it calls for automated interventions for reducing wastage of energy.

4.2.1   *Reducing Consumption by Preventing Wastage*

In many buildings it is not uncommon for Heating, Ventilation and Air-Conditioning (HVAC) systems to be turned ON at the beginning of the day only to be turned OFF at the end of the day. Also, people switch ON the HVAC system in a space at the start of a meeting; one would expect them to switch them OFF at the end of the meeting, but the latter rarely happens. Clearly, there is scope for automated means to manage the HVACs to reduce energy consumption and this must be exploited especially since HVACs constitute a fairly large proportion of energy consumed by buildings.

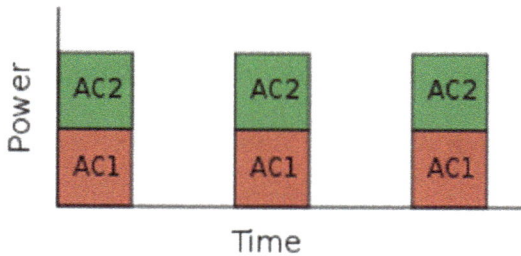

Fig. 4.3   High peak demand due to un-coordinated scheduling.

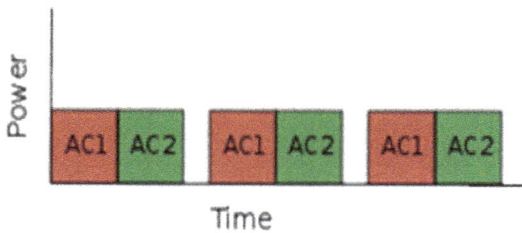

Fig. 4.4   Reduced peak demand under coordinated scheduling.

Switching of HVACs during the period of non-occupancy can reduce wastage. However, providing a low cost solution with minimum payback period for detecting occupancy with high accuracy is a challenge.

Further, pre-cooling a meeting space for an arbitrary period, which can go to half an hour prior to the scheduling time, is common. It requires analysis to determine the right pre-cooling period for a given space to prevent wastage of energy.

4.2.2   *Reducing Peak Demand*

Building electricity loads due to HVACs can contribute to the peak demand significantly. Consider a case of an office room fitted with 2 ACs. The usual practice is to switch on both the ACs immediately after entering the room. Now, both the ACs work for some time, the temperature of the room comes down and then both the ACs go off. Subsequently, both the ACs start again, when the room temperature goes high and the process continues. It can be observed from Figure 4.3 that under uncoordinated operation of ACs, the peak power requirement is that due to two ACs. But, if the ON-time of the ACs are staggered, the peak power requirement can come down to 50%, as shown in Figure 4.4. But, we must ensure that the desirable thermal comfort of the consumers is maintained while scheduling ACs.

Reduced and flattened peaks in power demand help utilities reduce investment in costly generating stations (oil/gas turbine and hydro) for meeting peak power requirements. This in turn, reduces the penalties (very high electricity tariff) levied by utilities on consumers for peak power consumption.

Therefore, we need techniques for automated intervention that will reduce peak demand while ensuring the desired thermal comfort.

4.2.3   *Improving Thermal Comfort*

Quite often, people complain of over-cooling and under-cooling, even in spaces with a good HVAC system. In most of the cases, this is not due to individual thermal preferences, but due to non-uniform cooling, especially in large spaces. Uniform cooling depends on the architecture of the space and the location of the HVAC outlets (e.g., IDUs). Therefore, it is challenging to improve the satisfaction levels of occupants in a space already fitted with HVACs.

In order to address these challenges, a holistic and systematic approach to provide thermal comfort in buildings is necessary in order to seek answer to the following fundamental questions.

How much energy is used by a given space?
How much savings will accrue from a well-designed HVAC strategy?
Can we improve the thermal experience of the individuals instead of providing an average level of satisfaction?
The answers to these questions depend on:

- The nature of the space — its size, shape, materials used, etc.
- The ambient conditions of the surroundings — outside temperature, humidity, etc.
- The desired "climate condition" for the space — inside temperature, humidity, etc.
- The energy consumption profile of each appliance in the space — which in turn may depend on how the users of that space utilize the appliances and set appliance specific parameters.
- Controllability that we can offer at the individual level to improve one's thermal experience.

Most of the commercial buildings today are equipped with air-conditioning resources. However, it still remains a fact that often the existing system fails to offer desired thermal comfort level. For example, it is not uncommon to see people come to over-cooled auditorium wearing jackets. Further, it is common in many buildings to turn on Heating, Ventilation and Air-Conditioning (HVAC) equipment early enough to pre-cool the space in order to keep it ready for the events like meetings. Often an arbitrary pre-cooling period is practiced, typically half an hour. In case of large spaces like auditoriums, a pre-cooling period of one hour is also observed. This practice often results in wastage of energy as it does not actually require so much time to bring down (in case of cooling) the temperature of an unoccupied space to the desired value.

> **Should HVAC appliances be switched off under no occupancy?**
>
> The quick answer is yes as it appears prudent to do so, if we care to save energy. But, if the room is highly likely to be occupied soon, won't it be wise to keep air-conditioners running during the period of non-occupancy, which we refer to as *unoccupied-period*? Clearly, the answer to the original question is not so clear cut.

Each space has some thermal capacitance and therefore it takes some finite time for the temperature to change. So, can we save energy by shutting OFF HVAC systems a little before the end of a meeting? How early can this be done without compromising thermal comfort?

We will see how the data driven model discussed in Section 4.5 can be used to inform us

- Whether we should start pre-cooling a given space to keep it climate-ready in time for a meeting, and if so, when?
- Whether we should keep a room cool during the *unoccupied-period* between two successive meetings?
- Whether we should switch off the air-conditioner before the end of a meeting so that energy can be saved, and, if so, when?

It requires analysis and its validation to find answers to these questions above. As we will see later in this section, the answers are not clear cut. When applied to a study space, the answers to these questions turned out to be: Yes, Maybe, and No!

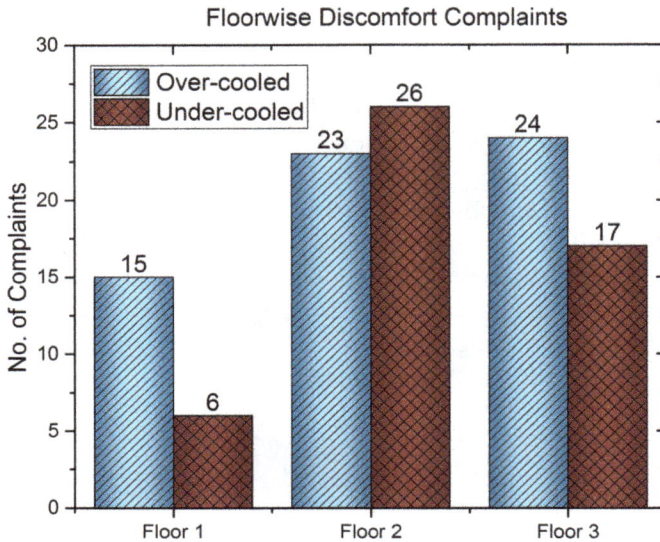

Fig. 4.5   Cooling related complaints in a commercial building.

It has been already discussed that despite adequate provisioning HVACs, energy-efficient and satisfactory thermal comfort is still a major problem. It has been observed that often people inside the same enclosed space like an auditorium complain of over-cooling and under-cooling at the same time. This is more due to the non-uniform cooling than individual thermal preferences. As an example, the number of complaints received per floor by the facility managers on a particular day, in an IT firm, is shown in Figure 4.5. These complaints are representative of the many defects in the present-day thermal comfort maintenance protocols implemented in buildings, which tend to over-cool or under-cool spaces, thereby affecting occupants' satisfaction adversely.

Typically in any facility, we find that spaces of different sizes and geometry are used for different purposes. Thermal comfort is affected by the thermal conditioning resources (HVACS, fans and windows) and architectural factors (size and the geometry of the space). Further, the architecture of a space plays a role in the choice of locations of the front-end components (viz., IDUs, ducts and VAV units) of a HVAC system. These affect the resulting thermal characteristics of a space and hence the thermal comfort of the users. Finally, an important factor identified above that affects thermal comfort, is non-uniform conditioning. Mitigating this problem needs a better understanding of the thermal characteristics of the space and the dynamics of air circulation in the space.

However, most often spaces are designed without energy conscious maintenance of thermal comfort in mind. This, in combination with *blind* protocols for HVAC operations and improper placement of sensors, often result in undesirable phenomena such as over-cooling, the presence of thermal gradients in large spaces like auditoria and big classrooms, if only HVAC is used, and wastage of energy due to pre-cooling for an arbitrary period.

Motivated by these multi-faceted considerations, this chapter offers a detailed discussion on how to achieve thermal comfort using a holistic approach. It begins with the factors influencing thermal comfort and the necessary interventions (manual and automatic) in various stages of providing the desired habitability. This is followed by discussions on the elementary principles behind conditioning of a space, its thermal characteristics, data driven modeling of heat transfer in building, its validation and a number of case studies establishing the need and effectiveness of a systematic approach in proving thermal comfort. Demand-Response (D-R) and the techniques used for maintaining thermal comfort while facilitating D-R control under power constraints (imposed by the utilities) are discussed at the end.

4.3 A Holistic Approach to Climate Control

Let us first examine the factors influencing the quality of thermal conditioning of spaces in a building and then arrive at the sequence of stages involved in providing thermal comfort. Also, we examine undesirable phenomena that can arise due to existing architecture and thermal conditioning resources, if no corrective action is taken and left to built-in controllers of the individual HVACs. Once equipped with this information, the possible interventions that can be made to meet the thermal comfort requirements are discussed at the end of this section.

4.3.1 *Factors Influencing Thermal Comfort*

Factors Influencing the Quality of Thermal Conditioning of Spaces in a Building

- ***Architectural features*** (viz., size of the space, location of the walls with respect to exposure to direct sunlight, floor types — flat/slanted, range of ambient temperature and humidity that the building caters to):

  The architecture of a space determines the maximum number of occupants, the ventilation provided and the type of usage possible (for example, a large auditorium would require a slanted floor), apart from relevant and useful information such as size of the space, the type of floor/walls, airflow in the space, etc.
- ***Conditioning resources*** (viz., HVACs, fans and windows and their numbers):
  Based on the architectural factors like size, exposure of the walls to direct sunlight, etc., and thermal comfort requirement, design decisions are taken on the number, type and the capacity of the resources to be provided for achieving thermal comfort. The resources can be HVACs, fans, open-able windows or any combinations of them.
- ***Determinants of thermal conditioning load*** (ambient conditions, desired conditions, occupancy):
  Climatic conditions, such as ambient temperature and humidity level, and occupancy level and locations affect the actions of the climate control system. Automatic interventions are needed in response to (changes in) these factors for maintaining thermal comfort and saving energy.
- ***Dynamics of the conditioning*** (run-time control to ensure that conditioning needs are satisfied):
  Run-time interventions are necessary to maintain temperature within an acceptable band that accommodates variations in occupancy/ambient conditions.

The architecture and the prevalent external conditions determine and constrain the thermal conditioning requirements of a space. This conditioning is achieved by examining one or more possible resources and making appropriate choices during the *Thermal Design stage.*

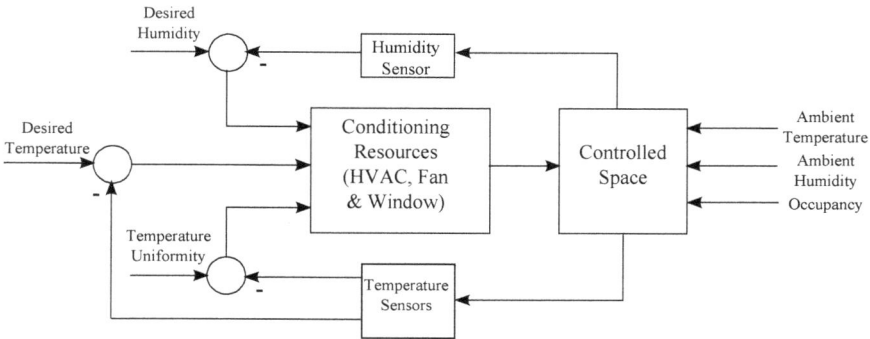

Fig. 4.6   Climate control of a space.

Using the specifications for the desired thermal conditions and considering the extant temperature and humidity along with the details of the usage/users, the conditioning resources chosen are *initialized*. This could involve switching on HVAC appliances, fans, or even opening of windows for natural ventilation. Initialization depends on the available resources, user requirements and thermal disturbances. For example, the number of HVACs and/or fans to be switched on before a meeting starts, is based on the expected occupancy, desired temperature of the users and ambient conditions.

It is the system's responsibility, through appropriate *run time control*, to adjust and manage the conditioning resources, so that the space is thermally comfortable.

A scheme for automatic run-time control is depicted in Figure 4.6 which shows that control action on the thermal conditioning resources (available for a given space) is taken based on the i) desired temperature, ii) desired humidity and iii) (non-)uniformity of cooling termed as *thermal gradient* (prominent in large spaces like auditorium — discussed in Section 4.8). It can also be observed from Figure 4.6 that in order to maintain the desired thermal comfort level, feedback of temperature and humidity is taken to accomodate the variations in heat load and the ambient conditions.

4.3.2   *Stages Involved in Providing Thermal Comfort*

Given the factors mentioned above, the following stages are involved in the process of providing thermal comfort in a space.

- Architectural Design

- Thermal Design
- Initialization of conditioning resources
- Run-time control of conditioning resources

The architectural design and thermal design stages are one-time decisions for a particular space. On the other hand, the climate control actions during initialization are done once for each "use" of the space, and run-time control is continuous and dynamic due to the variable nature of the inputs to these stages.

These last two stages depend to a large extent on the geometry of the space, the details of airflow within the space and the occupancy details. These factors determine the energy needed for cooling and the time it takes to cool a particular part of the space. Hence, the details of these stages are determined through empirical observations which inform the decisions pertinent to the two stages. The decisions relate to which parts of the space need to be conditioned and when the conditioning should start so that the occupants will be comfortable when they are seated in a particular part of the space.

As the practice goes, the thermal conditioning resources are deployed in a space considering the worst-case scenario of maximum occupancy and the most unfavorable ambient conditions that can be expected in a particular geographic region. This is done keeping in mind the consumer's comfort under all possible scenarios. It is often observed that, irrespective of heat loads (occupancy) and ambient conditions, the full capacity of the conditioning resources is usually switched on. For example, it is usually the case that irrespective of occupancy level, all the available air-conditioner units in an auditorium are switched on leading to over-cooling. Similar observation is made in large classrooms in academic building, where all the in-door units (IDU) are switched on (owing to scattered sitting) leading to over-cooling, till someone takes the effort to switch off some of the IDUs.

Further, *thermal gradient* is observed in spaces due to the large size and impact of geometric constraints on the location of HVAC duct outlets/IDUs. This leads to uneven cooling within sub-spaces.

4.3.3   *Sensing Undesirable Phenomena in Spaces*

We will now discuss the undesirable scenarios pertaining to thermal comfort that can arise and the necessary interventions that can take care of the wastage of energy and provide desired thermal conditioning of a space.

The phenomena which must be avoided to improve thermal comfort include:

- **_Thermal Gradient:_** The location of duct outlets/IDUs plays an important role in how the temperature distribution happens in the given space. Often, due to ad-hoc geometric locations of the duct outlets, non-uniform distribution of temperatures is observed in the spaces with 1 to 2°C difference. This difference results in the creation of hot and cold subspaces.
- **_Nonuniform Temperature Distribution:_** Any window/split AC unit often creates thermal dissonance as the velocity of the chilled air thrown from its evaporator reduces as it travels across the closed space. This causes variation in the temperatures across the space thus providing different comfort levels in the sub-spaces.
- **_Over-cooling and Under-Cooling:_** Quite often, not much thought is given while operating HVACs — how many ACs to use, for how long, etc. This unchecked usage manifests into over-cooling or under-cooling leading to inefficient energy usage and inferior thermal comfort. It may also happen due to system's temperature control algorithm gone awry.
- **_High Internal Humidity:_** It is known [Fanger (1973)] that high humidity causes discomfort to occupants in a given space. In spaces with operable window resources, opening windows to allow passage to low humid outside air into the space improves the state of comfort.

Real-time information about the current spatial and temporal conditions are available in modern buildings through IoT Systems. Data from these systems can be exploited to provide better climate control of spaces. Some possible interventions are outlined in the next section.

4.3.4 *Analyzing Possible Pro-active and Reactive Interventions*

Now we examine a set of possible interventions. Some of these require the design phase to anticipate their use at run-time and provide for them at design time.

- *Smart Seating Arrangement — Cooler Seats First Policy:* Thermal gradient data helps us identify cool and warm sub-spaces within a space. Displaying a thermal map of the seats and asking people to occupy seats based on their thermal preferences, will improve their comfort.
- *Provide Fans to Improve Air Circulation:* Use of fans improves air circulation in all the sub-spaces leading to uniform temperature within the space. This gives flexibility to the occupant to sit anywhere in the space and feel comfortable.
- *Zone Based Cooling:* The location of IDUs affects the cooling rates in different parts of the space — some parts cool faster than others, thereby dividing the whole space into zones which can be put to use individually.
- *Seating Arrangement — One Zone At a Time Policy:* Since it is known that thermal behavior in zones vary, asking people to occupy colder zones completely before opening up other zones is a suitable intervention for better thermal comfort guarantees.
- *Determine Cooling Times:* Knowing precisely the time required to achieve a state of comfort with different number of ACs help avoid over-cooling situations. Thermal models from existing work like [Karmakar *et al.* (2015a)] can be utilized to determine the cooling time of every sub-space.
- *Adjust Number of AC units used:* The number of ACs within a space has a distinct affect on factors such as the area being cooled, the time required to cool, etc. Knowing their behavior helps in choosing ACs for an event.
- *Using Custom Temperature Controller:* In cases where relying upon system's temperature controller results in over-cooling, custom temperature control algorithm is to be used to maintain the desired set point or band.

Before planning or employing any interventions towards efficient climate control, it is necessary to understand the thermal dynamics of the space. This requires thermal modeling of a building space.

4.4   Thermal Modeling of Building

Thermal model of a building is important for its use in energy management and providing thermal comfort. This is because the control of thermal environment in many of its forms depends on the model of the heat dynamics in a building. Therefore, the quality of the thermal model is also important. For an effective and implementable control, the requirements can be contradictory. Firstly, the model must be accurate so that it generates correct control output(s) for particular input(s). Secondly, it is desirable that the model is computationally simple so that the control system can be implemented using a low cost computer and at the same offering the desired real-time response. This is important as any energy management solution must be economically attractive to the consumers with minimum pay-back period. Usually, the more accurate a model, the more it becomes complex and therefore computationally intensive and costly.

The approaches on thermal modeling can be broadly categorized as

  i) first principles of thermodynamics (FPT),
 ii) data-driven, and
iii) hybrid.

In addition, another important approach is modeling the system (involving thermal conduction) using the mathematically analogous system of electrical network of resistance (R) and capacitance (C) or RC network [Davies (1983); Fraisse *et al.* (2002)]. Hence, we consider this approach under FPT. The RC modeling is preferred by control system engineers experienced in the analysis of electrical circuits.

However, it is difficult to find a single modeling approach, which can be gainfully exploited in real-world applications. The primary reason behind this is the complexity involved in developing an accurate model, which also captures the measurement inaccuracies, stochasticity of inputs like solar heat through the windows and walls when exposed to direct sunlight. This also makes the case for hybrid modeling approach. Further, it is highly difficult to capture the variations in the parameters like thermal capacity of a building wall, heat loss/gain due to thermal conduction that may change over time.

4.4.1 *First Principles of Thermodynamics (FPT) and Electrical Analogy*

FPT approach is based on the physics of heat transfer and thermodynamics involving mathematics, which describes the heat dynamics of a building. As discussed, the RC model is also based on the first principles developed on the mathematical analogy between heat dynamics and flow of electricity through the network of resistances and capacitances. In RC model, capacitance represents the thermal capacity and resistance represents the thermal resistance. Note that thermal capacitance is the physical property of a material/object that describes its capability to store heat energy, while thermal resistance describes the property of the material/object to resist flow of heat through it. Also, it may be recalled from elementary principles of physics that it is the difference in temperature that causes flow/exchange of heat between two bodies/objects. This is analogous to the flow of current (rate of flow of charge) that occurs when there is difference in potential between two points in a conducting material having some resistance. Thus, in RC model current equals heat flow rate and voltage (potential) equals temperature.

The RC model has a number of advantages. The model is

  i) solvable using electrical engineering rules,
 ii) flexible enough to model large buildings, and
iii) suitable for hybrid approach, where parameters can be derived from data.

However, in spite of the several advantages of RC modeling, the major disadvantage is that for a real-world building the complexity of the RC network grows very large. This involves larger computational time and effort in model development and can be a deterrent to develop a low cost controller for any application like providing the desired thermal comfort with reduced consumption and/or under peak demand constraint. A detailed discussion on the electrical analogs of thermal systems and RC modeling is presented in Appendix C.2.

In this context, it is important to note that in developing thermal model of a building from the first principles, usually the following assumptions are made [Skruch (2015)].

(1) The thermal properties of the building materials used in the building construction are known.
(2) Indoor air temperature is the same at any point in the room or the zone.
(3) The air-density is constant in all the rooms/zones in the building.
(4) It is ensured by the ventilation system that the amount of air that is removed is replaced by the same amount.

4.4.2 Data-Driven Approach

A data-driven model is developed based on the measured field data obtained by conducting experiments in real-world system, a building or a room in this case. The data are analyzed by various algorithms to generate a model of the system.

Data-driven modeling is a black-box approach, where the functional models are developed based on measured data considering the system as black-box. The availability of huge amount of data from millions of smart meter and thermostat installations along with weather data makes the data-driven approach an attractive choice for the control engineers [Jain *et al.* (2017)]. Artificial neural networks (ANN) have been used for building thermal model of buildings [Huang *et al.* (2013); Dong *et al.* (2008)], based on available data.

However, in data-driven approaches, due to the absence of physical representation, the internal dynamics of the system acts as a black-box and can be vulnerable to poor choice of data and uncertainty in datasets such as weather forecast data. Also, the computationally intensive nature of this approach is often a deterrent to the development of practical and low-cost controllers.

4.4.3 Hybrid Modeling

Hybrid modeling is a combination of modeling from first-principle and empirical data. This is a middle path approach, where a simpler FPT model is used to create a basic structure of the model and the parameters of the FPT is estimated and/or refined using the available data [Goyal *et al.* (2011)]. Thus, the hybrid approach offers a computationally simpler model with good accuracy.

4.4.4 Practical Limitations

One of the fundamental assumption of the building thermal modeling techniques [Bueno *et al.* (2012); Skruch (2015)] is that HVAC system is ideal and work at constant efficiency irrespective of the disturbances like changes in the ambient temperature T_a and heat loads — body heat and heat generating sources like computers. However, in reality the efficiency of a HVAC system varies widely and dynamically with changes in the ambient temperature. For example, in the case of an air-conditioner (AC), the effect of change in T_a is considerable on the rate of heat transfer at the condenser coil in the Out-Door Unit (ODU) and therefore on its efficiency.

Further, the following observations also deserve consideration in the modeling of closed spaces in a building.

i) The effect of changes in T_a on the inside air temperature, is sluggish due to the thermal capacity of the walls.
ii) The power consumption of an AC depends also on the set/desired temperature T^s. This is due to the fact that the rate of heat transfer is proportional to the difference in temperature between the two bodies (the evaporator coil in the In-Door-Unit (IDU) of an AC and the room air).

Therefore, the power requirement of an AC varies with the variations of both the ambient temperature T_a and the set temperature T^s. These observations suggest that a practical thermal modeling of a building space, should be adaptive.

4.5 Adaptive Hybrid Modeling Approach

The adaptive approach is similar to the hybrid model, where we model the HVAC system from the first principle, estimate the characteristic constants (parameters) from measured data, but keep updating the parameters online in order to adapt to the disturbances like changes in T_a and heat loads.

The key idea behind this adaptive approach in thermal modeling of a closed space is as follows [Karmakar *et al.* (2018)]. Considering the fact that the effect of changes in ambient temperature T_a is observed in corresponding changes in the inside wall surface temperature T_{wi}, it is the convective heat transfer between the inside wall and the room-air that affects the thermal environment (due to changes in T_a) of the closed space. In other words, so far as the thermal conditioning of a space is concerned, the inside walls can be considered as variable heat sources, like human body heat (depends on the number of occupants at a given time). When one simply models the HVAC system and updates its parameter based on the changes in T_{wi} and heat loads, a simple, less computationally intensive model requiring minimum measurement (viz., T_a, T_{wi}, and T_i) can be developed, which does not assume constant efficiency of HVAC system unlike other models.

The main assumptions in the adaptive approach are the following.

- The effect of T_a on the internal temperature T_i is much slower compared to its effect seen by the HVAC system (to be precise, the condenser coil in the ODU of the AC). This is true in all/most cases, except when the walls have very low thermal resistances and thermal capacitances.

- Disturbances like changes in ambient temperature T_a are not so very rapid so as to cause large changes in the characteristic constants (parameters) within a short period. Therefore, it is feasible to update these parameters based on the immediate historical data, say from previous 15 minutes and this assumption is reasonable.

The advantages of this approach are as follows.

- It requires only a model of HVAC system based on first principles, which is simple and therefore less computationally intensive to run on controllers based on low cost computers like Raspberry-Pi,
- It does not assume constant efficiency of HAVC system unlike other models like RC network,
- Data-driven parameter estimation makes the model widely applicable to varied types of closed spaces without requiring specific information about the materials of the walls, their number of layers, thickness, etc., and
- The model adapts to the disturbances, including unpredictable ones like changes in weather and take appropriate control action to maintain the desired thermal comfort.

The adaptive model is intuitive and simple. Using this, it is also possible to predict power and energy demand so that appropriate demand-Response (D-R) control action can be taken under peak demand constraints. Further, this approach is capable of adapting to the changes in physical parameters of walls like thermal resistances and capacitances, which may occur over time.

Let us take the example of a space, where the temperature is maintained by an air-conditioner (AC). Note that the technique developed for modeling a space cooled by AC will be also applicable where the requirement is heating a space using a heater for maintaining thermal comfort. This is because of the fact that both heating and cooling follow the same principles of heat transfer and thermodynamics.

In the following section first, the working of an AC and the development of the system model from the first principles of thermodynamics is discussed. This is followed by its implementation and results in managing energy and thermal conditioning of a space in general.

4.5.1   *Notations*

E	Energy Consumption Rate
W	Work Done
h	Coefficient of Heat Transfer
T_i/T_z	Inside/Zone Air Temperature
T_s	Set Temperature Value
T_a	Ambient Temperature
T_{wi}	Surface Temperature of inner walls
T^E	Temperature of Evaporator Coil
T^C	Temperature of Condenser Coil
Q	Total Heat Energy Transfer
Q_{hi}	Overall Heat input rate from various sources
Q_{hw}	Convective Heat Transfer through the walls
Q_C	Rate of Heat Energy released through external fan coil in AC ODU
Q_E	Rate of Heat Energy absorbed through Evaporator coil in AC IDU
Θ_i	Thermal Capacity of the space Z_i

4.5.2   *System Model*

Consider a simple system of a closed space (a room/zone) in a building, where the thermal comfort is maintained by a window air-conditioner (AC). The basic assumption for the system is that the maintenance of thermal comfort is limited to maintaining the inside temperature T_i, i.e., other parameters like humidity are not taken into consideration. This is due to the fact that while maintaining temperature to a set value T_s, an AC also maintains humidity within the desired level by default.

From the point of view of maintaining thermal comfort in a space, i.e., from control point of view, the changes in the ambient temperature T_a and heat loads are disturbances that affect the inside temperature T_i. Note that the variation in T_a, in effect, changes the surface temperature of the inner walls T_{wi}, which acts as heat source, as discussed in Section 4.5.

Now, since the effect of T_a on T_{wi} is sluggish due to the thermal capacity of the wall, the model of AC can be adapted by updating the characteristic constants (parameters) of the model of AC based on the measured data. This adaptation can be made from archival data or from the measured data from the immediate past — about 15 minutes.

Using this model, the goal is to develop an effective strategy of HVAC appliance control in order to save energy while maintaining the desired thermal comfort. To achieve this goal it is necessary to i) analyze the

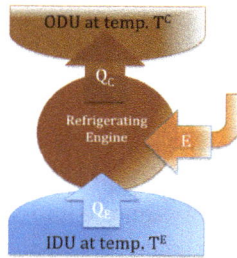

Fig. 4.7   Energy flow in an air-conditioner.

relationship between energy consumption rate E at a given set temperature T^s, its dependency on ambient temperature T_a and heat loads, and ii) estimate energy demand.

4.5.3   *Thermal Modeling of a Building Space*

Let us now look into the energy flow within an AC (a HVAC system) and develop a fairly simple thermal model of a space cooled by it. Empirical validation of the model and a technique for the prediction of power and energy demand will also be discussed. This simplified model of the AC system is based on the fundamental principles of thermodynamics and establishes the relationships between energy consumption E (also referred to as the work done W, by the compressor), the outdoor (ambient) temperature T_a and inside/zone air temperature T_i.

Heat transfer takes place i) between the internal cooling coil and the room air when they come in contact as facilitated by the fan (in the evaporator) and ii) between the external condenser coil and the outdoor air coming into contact forced by the outdoor fan (in the condenser). Figure 4.7 shows the energy flow within an AC system. Let W denote the work done (energy consumed E) by the compressor in the ODU (Out-Door Unit) of the AC, which heats up the refrigerant. The refrigerant flows through the external fan coil and releases heat Q_C to the atmosphere. The high pressure refrigerant then gets cooled by adiabatic expansion (through a throttle valve) and passes through internal coil (evaporator) in IDU (In-Door Unit) of the AC. The cold refrigerant in the evaporator absorbs heat Q_E from the air (force-circulated using a fan in IDU) in the space for which it maintains thermal comfort.

The rate of work done by the compressor W, i.e., the power (*watt*) requirement of the compressor can be calculated from the first law for a cyclic

thermodynamic process [Young and Freedman (2008)], and it is governed
by the following general equation

$$W = Q_E + Q_C \tag{4.1}$$

where, Q_E Joules/sec (J/s) is the rate of heat energy absorbed by the
evaporator coil, i.e., the energy that goes into the refrigerating engine. Note
that heat transfer takes place between the internal fan coil (evaporator) and
the room air when they come in contact as facilitated by the fan. Therefore,

$$Q_E = h_1(T_i - T^E) \tag{4.2}$$

where, T_i is the space temperature, T^E is the temperature of the evaporator coil and h_1,[3] is the overall heat transfer coefficient. Further, note
that in this simplified model, it is assumed that the design parameter T^E
(temperature of the evaporator coil) remains constant. This assumption is
reasonable considering the fact that the bulk of the heat exchange that takes
place in the evaporator is in the form of latent heat due to phase change
(liquid to gas) of the refrigerant. However, it is worth noting here that if
T_a changes, it affects the heat transfer in the condenser and thereby the
temperature of the compressed refrigerant and T^C. Implication of changes
in T_a is discussed in Section 4.5.3.1.

Q_C is the rate of heat energy released by the condenser to the environment:

$$Q_C = h_2(T^C - T_a) \tag{4.3}$$

where, T^C is the condenser coil temperature, T_a is the ambient temperature
and h_2 is the overall heat transfer coefficient.[4]

In a refrigeration cycle (applicable in the case of AC), work goes into
the system (refrigerating engine) and therefore, W is negative. Since Q_E is
the heat energy that goes into the refrigerating engine and Q_C is the heat
energy that goes out of it, Q_E is positive and Q_C is negative. It follows
from Equations (4.2) and (4.3) that

$$W = -h_1(T_i - T^E) + h_2(T^C - T_a) \tag{4.4}$$

Equation (4.4) shows that the energy consumed by the HVAC system
depends both on the set temperature $T^s(T_i = T^s$ at steady state) for the
space that it tries to maintain and the ambient temperature T_a. However,

---

[3] $h_1 = hA_1$ where, h [$Joules/(m^2.K.sec)$] is the coefficient of heat transfer of the
refrigerant and $A_1[m^2]$ is surface area of the evaporator coil.
[4] $h_2 = hA_2$ where, h [$Joules/(m^2.°C.sec)$] is the coefficient of heat transfer and A_2 is
the surface area of the condenser coil.

there is an apparent contradiction in the energy balance as formulated in Equation (4.4) that if ambient T_a increases, the second quantity in the RHS decreases as $T^C > T_a$ and therefore, W decreases with increase in T_a. But, it does not happen so, because change in T_a also causes a corresponding change in T^C, which we will discuss in detail in Section 4.5.3.2.

4.5.3.1 *Modeling Building Space Cooled by an AC*

Let us first carry out a simplified theoretical analysis of cooling a space by an AC based on the fundamental principle of Thermodynamics. Followed by this analysis, we will look into how this simple model can be useful for real-life application when refined with experimental data to generate the thermal characteristic constants.

The assumptions made in developing this model are — i) all the heat sources (viz., human body, heat producing equipment, conductive heat through the walls and ii) heat input from the outside environment due to ventilation/leakage paths) are constant. Later, we will see how this model can be adapted to the changes in the ambient T_a and occupancy. Only the variations in T_a and occupancy are considered, for it is reasonable to assume that heat input through leakage paths and from the heat producing equipment like computers do not vary.

In a building space, let Q_{hi} denote the overall heat input rate from the various sources, some of which are mentioned above. We can write,

$$Q_{hi} = Q_h + Q_{hw}$$

where, Q_{hw} denotes the convective heat transfer through the walls, Q_h denotes the heat input from other sources. For simplicity of exposition, let us first take a case assuming heat load due to occupancy as constant at a point of time when we are analyzing the effect of changes in T_a.

$$Q_{hi} = Q_h + h_w(T_{wi} - T_i) \tag{4.5}$$

where, h_w is the overall convective heat transfer constant and T_{wi} is the temperature of the inside of the walls assuming all the walls have same temperature and it is uniform throughout the surface.

The heat taken away by the evaporator coil in the IDU is $h_1(T_i - T^E)$ (Equation (4.2)). Therefore, the total heat energy transfer Q that takes place within a space is

$$Q = Q_h + h_w(T_{wi} - T_i) - h_1(T_i - T^E) \tag{4.6}$$

Now, the same transfer of energy is what causes change in the room temperature and this can be defined by

$$Q = \Theta_i \frac{dT_i}{dt} \tag{4.7}$$

where Θ_i denotes the thermal capacity of the space Z_i.

Therefore, it follows from Equations (4.6) and (4.7) that

$$\Theta_i \frac{dT_i}{dt} = Q_h + h_w.T_{wi} + h_1.T^E - T_i(h_w + h_1)$$

Case 1: No Change in T_a

When there is no change in T_a, T_{wi} remains constant and this is valid for any time duration, till T_a changes [Karmakar *et al.* (2018)]. In such a case, we can write,

$$\frac{dT_i}{dt} = a - bT_i \tag{4.8}$$

where,

$$a = \frac{Q_h + h_w.T_{wi} + h_1 T^E}{\Theta_i} \quad \text{and} \quad b = \frac{h_w + h_1}{\Theta_i} \tag{4.9}$$

Solving Equation (4.8), with the initial conditions that $t = 0$, $T_i = T^H$, we get

$$t = -\frac{1}{b} ln \frac{a - bT_i}{a - bT^H} \tag{4.10}$$

Therefore, we can obtain the thermal profile of an AC cooling a space from a temperature T^H as

$$T_i = p + qe^{-r \times t} \tag{4.11}$$

where, $p = \frac{a}{b}$, $q = \frac{bT^H - a}{b}$, and $r = b$.

It follows that Equation (4.10) can be rewritten as

$$-\frac{1}{r} ln \frac{T^H - p}{T_i - p} \tag{4.12}$$

Equation (4.12) can tell us how much time an AC will take to bring down the temperature of a space from T^H to any value T_i given that the occupancy and the T_a remains unchanged. However, it is to be remembered that any change in T_a will finally result in a change in the characteristic constants p, q and r of the thermal profile. We will now analyze how this simpler model can be made adaptable to take care of the more realistic case of variable T_a in the following section.

Case 2: Variation in T_a

This is a more general case when we consider the variations in ambient temperature T_a. As discussed earlier that due to conductive heat transfer through walls, variations in T_a cause changes in surface temperature T_{wi} of inside walls, which act as a heat source. The conventional approach is capturing the effect of T_a on the inside temperature of a room is by thermal modeling of the walls, as discussed in Appendix C.2.2. It not only increases the complexity of the model, increasing computation time requirement, but also requires the information about materials of construction of each layer of the walls in order to derive the correct thermal resistances and capacitances involved in the model. Further, it is also a fact that thermal properties of a wall material can change over time, which cannot be captured in the conventional model.

A simpler solution is presented here. The key idea behind this approach is the fact that for all practical purposes, a change in T_a, which is less than $0.5°C$, does not cause any appreciable change in thermal comfort to the consumers inside a building space. Therefore, one need not keep track of every small change in T_a in order to update the thermal model for analyzing its effect on the inside temperature T_i and taking any control action necessary to provide thermal comfort.

Since the effect of changes in T_a results in changes in the characteristic constants of the thermal profile and it is practical to discretize the changes in T_a with a step size of $0.5°C$ so that one only needs to generate a small lookup table for the values of characteristic constants p, q and r for a small and finite set of different values of T_a covering the entire range that a building in a particular geographical region can see. Thus this model adapts to variations in T_a simply by using an applicable set of values of the characteristic constants generated by regression technique, either by using the archival thermal profile or by learning from data measured during the recent past.

4.5.3.2 Estimation of Characteristic Constants

The characteristic constants of the thermal profile are estimated using data from experiments in real-world spaces. Experiments can be carried out by running air-conditioners (AC) to measure the changes in temperature of any closed space under various ambient temperatures and the thermal profile (zone temperature T_z versus time t) can be obtained using regression technique to fit the basic structure developed from the first principles

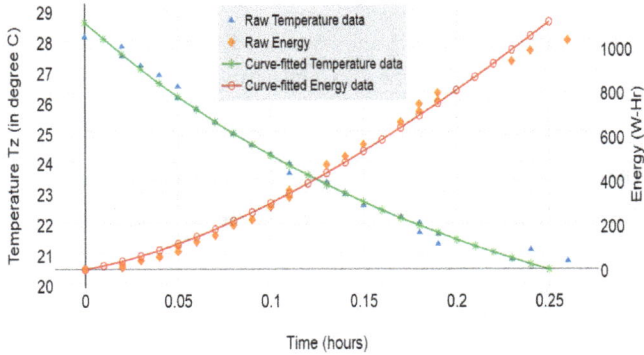

Fig. 4.8   Thermal profile of cooling a classroom and the corresponding energy consumption to cool to a particular temperature (at $T_a = 31^{\deg}C$).

(Equation (4.11)). The example of the thermal profile of a real-world classroom is shown in Figure 4.8.

4.5.4   *Effect of Changes in Ambient Temperature T_a on Energy Consumption*

According to the second law of thermodynamics for a cyclic process [Saha and Srivastava (1967); Rogers and Mayhew (1982)], we can write

$$\frac{Q_E}{T^E} = \frac{Q_C}{T^C} \qquad (4.13)$$

In order to analyze the effect of T_a on energy consumption, we consider that the system is in steady state with $T_s = T_i$, and therefore, Q_E/T^E remains constant.

It follows from Equation (4.13) that

$$\frac{Q_C}{T^C} = constant$$

Substituting the value of Q_C from Equation (4.3), we get

$$\frac{h_2(T^C - T_a)}{T^C} = h_2\left(1 - \frac{T_a}{T^C}\right) \Rightarrow \frac{T_a}{T^C} = constant, \text{ as } h_2 \text{ is constant.}$$

Let us assume that the ambient temperature changes to $T_a + \delta$. It will result in a corresponding change in T^C by γ so that

$$\frac{T_a + \delta}{T^C + \gamma} = constant$$

Considering the fact that T^C is always greater than T_a, we can write

$$\gamma > \delta \qquad (4.14)$$

Let W_2 denote the energy consumption rate, when ambient temperature changes from T_a to $T_s + \delta$. From Equation (4.4), we can calculate the change in energy demand rate ΔE as

$$W_2 - W_1 = h_2[(T^C + \gamma) - (T_a + \delta)] - h_2(T^C - T_a) = h_2(\gamma - \delta) \qquad (4.15)$$

where, W_1 denotes the energy consumption rate corresponding to the original ambient temperature T_a.

It shows that energy consumption rate is directly proportional to the ambient temperature T_a, as $\gamma > \delta$ and both δ and γ are either positive or negative. It is concluded from Equation (4.4), that power consumption of an AC compressor is directly proportional to the changes in ambient temperature T_a.

4.5.5 *Effect of Changes in Set Temperature T_s on Energy Consumption*

ACs are designed such that the set temperature T_s is maintained irrespective of the value of ambient temperature T_a, subject to design limits.[5] This is guided by the user requirement that average zone temperature will be maintained at the desired T_s irrespective of the variations in T_a.

Consider the steady state when T_i reaches the set point, i.e., when $T_i = T_s$. The assumption in this case is that there is no change in ambient temperature T_a and therefore, Q_C is constant (Equation (4.3)). So, from Equation (4.13), one can write

$$\frac{Q_E}{T^E} = constant$$

Substituting the value of Q_E from Equation (4.2) and considering the fact that $T_i = T_s$ and h_1 is constant

$$\frac{h_1(T_s - T^E)}{T^E} = h_1\left(\frac{T_s}{T^E} - 1\right) \Rightarrow \frac{T_s}{T^E} = constant$$

Let us assume that the set temperature T_s is changed from T_s to $T_s + \alpha$. It will result in a corresponding change in T^E by β so that the ratio ($T_s +$

---

[5]For a particular zone volume and the estimated heat produced in the zone, an AC is designed to cool the zone to a lowest value of T_s considering the highest T_a than can be expected due to seasonal variations.

$\alpha)/(T^E + \beta)$ remains constant, according to Equation (4.13). [Note that at steady state $T_i = T_s$.]

$$\frac{T_s + \alpha}{T^E + \beta} = \frac{T_s}{T^E} \Rightarrow \frac{T_s}{T^E} = \frac{\alpha}{\beta} \qquad (4.16)$$

Considering the fact that $T_i(= T_s) > T^E$, it is trivial to state that $\alpha > \beta$.

Let W_1 denote the energy consumption rate when the set temperature was T_s and W_2 denotes the energy consumption rate, when the set temperature is changed to $T_s + \alpha$. At steady state, the zone temperature T_z becomes equal to T_s. By substituting T_z with T_s in Equation (4.4) we get W_1. Similarly, by substituting T_z with $T_s + \alpha$ and T^E with $T^E + \beta$, we get W_2. So, we calculate the change in energy consumption rate as follows.

$$W_1 = -h_1[T_s - T^E] + h_2(T^C - T_a) \qquad (4.17)$$

$$W_2 = -h_1[(T_s + \alpha) - (T^E + \beta)] + h_2(T^C - T_a) \qquad (4.18)$$

$$W_2 - W_1 = -h_1(\alpha - \beta) \qquad (4.19)$$

It shows that change in the rate of energy consumption is inversely proportional to the change in the set temperature T_s as $\alpha > \beta$ and the fact that α and β are either both positive or both negative according to Equation (4.16).

4.5.6   *Energy Consumed in Maintaining Thermal Comfort*

Power and Energy Demand:

Given a thermal profile of cooling, how do we calculate power and energy required to maintain the desired thermal comfort in a space?

The energy balance Equation (4.4) can help us, if we can formulate a method to derive power demand W of an air-conditioner (AC) as a function of time t. This will enable us to find the energy consumption within a time period $[0, t_b]$, when the AC brings down the zone temperature T_z from T^H at time $t = 0$ to T_s at $t = t_b$.

Substituting the value of T_z from Equation (4.11) in the energy balance Equation (4.4), we get the power demand W of an AC at any zone temperature $T_z = T_s$.

$$W = -h_1(p + qe^{-r \times t} - T^E) + h_2(T^C - T_a)$$

$$W = \phi - \psi e^{-r \times t} \qquad (4.20)$$

where, $\phi = h_1 T^E - h_1 p + h_2(T^C - T_a)$, $\psi = h_1 q$ and $r = \frac{h_1}{\Theta_i}$.

4.5.6.1 *Energy Consumption within a Period $[0, t_b]$*

Let t_b denote the *time required to cool a zone* from T^H to T_s and let us assume at $t = 0$, $T_z = T^H$.

We can compute the total energy E consumed during the period $[0, t_b]$ using Equation (4.20) as follows.

$$E = \int_0^{t_b} W dt = \int_0^{t_b} \phi dt - \psi \int_0^{t_b} e^{-r \times t} dt$$

$$E = \phi t_b - \frac{\psi}{r}(1 - e^{-r \times t_b}) \qquad (4.21)$$

4.5.6.2 *Predicting Energy Savings*

Often energy gets wasted due to the pre-cooling period being longer than what is required. Further, it is also common practice to keep HVACs running during the *unoccupied-period* between two successive meetings. Can we conclude quantitatively if it is beneficial to switch off or to keep a HVAC system running during the *unoccupied-period* of non-occupancy between two successive meetings?

Let us look into the specifics on i) how energy can be saved by choosing the pre-cooling period in a principled way, ii) how a quantitative assessment can be made if it is beneficial to keep an HVAC system running during the *unoccupied-period* and iii) whether we should shut-off ACs earlier than the end of meetings.

The characteristic constants of Equations (4.11) (cooling) and (4.21) (consumption) of a case study presented in Figure 4.8 are shown in Table 4.1. Using these characteristic constants, the power demand and the energy demand can be calculated.

• **Power Demand:** W^D of a HVAC, i.e., the wattage requirement at any instant to maintain set temperature T_s can be obtained from Equation (4.20) once we know the time t_b (from Equation (4.12)) it takes for the zone temperature T_z to reach T_s from T^H.

Table 4.1 Characteristic constants for temperature (cooling) and energy consumption profile of Figure 4.8 ($T_a = 31°C$).

Cooling (T^H to T_s)			Energy Consumption		
p	q	r	ϕ	ψ	r
16.7	11.9	4.59	8736	7194	4.59

• **Energy Demand:** $E^D (= W^D xt)$ required to bring down zone temperature from T^H to T_s can be calculated using Equations (4.12) and (4.21). Note that we calculate E^D in two steps — first, we find the time t_b to attain T_s from T^H (Equation (4.12)) and then we use the value of t_b to obtain E^D (Equation (4.21)).

4.5.7 *Responding with Data Driven Decisions: A Case Study*

Let us now discuss, along with the results of a case study, how data driven model can help in i) deciding pre-cooling time and arriving at a conclusion if cooling during *unoccupied-period* or shutting off HVACs early can be beneficial in saving energy.

4.5.7.1 *Pre-Cooling*

The pre-cooling period t_p is the time required to bring down initial zone temperature $T_z = T^H$ and the desired temperature $T_z = T_s$. The value of t_p can be calculated using Equation (4.12) once the thermal characteristic constants (p, q and r) are obtained from the look-up table for a particular T_a. The results of experiments to determine the various types of rooms with no occupancy under various ambient conditions show that average pre-cooling period is very low (~ 10 minutes due to non-occupancy during pre-cooling) as against the usual practice of 30 minutes to 1 hour for a typical classroom and a big auditorium. This is corroborated by analysis; the required pre-cooling period (to bring down T_z from $28.5°C$ to $23°C$) in case of a typical classroom under $T_a = 31°C$ can be obtained using Table 4.1 and Equation (4.12) as follows.

$$t_p = \frac{1}{4.59} ln \frac{28.5 - 16.7}{23.0 - 16.7} = 0.14 \text{ Hr.} = 8.4 \text{ min.}$$

4.5.8 *To Cool or Not To Cool during Unoccupied-Period*

Let us formulate the conditions under which it may be beneficial to run HVACs during the unoccupied-period t_g between two successive use of a space.

If a HVAC is allowed to run during an unoccupied-period t_g, it will consume energy, which can be wasteful. However, if the HVAC is switched off during the unoccupied-period t_g, the temperature of the zone will rise to T^H from the existing zone temperature $T_z (= T_s)$. Let W_m denote the power demand of the AC unit to maintain the temperature at T_z, if it is kept running. Now, in order not to compromise thermal comfort, a pre-cooling period t_p is also to be considered within the unoccupied-period. So, effectively we can switch-off the HVAC during the period $(t_g - t_p)$ only. Thus, the amount of energy that can be saved by switching off the HVAC will be

$$(t_g - t_p)W_m$$

where, W_m is the power demand of the AC to maintain the zone temperature at the set value. Therefore, it will be beneficial to run HVAC if the following condition is satisfied.

$$(t_g - t_p)W_m \le E_p$$

where, E_p is the energy that will be consumed for pre-cooling from T^H to the desired set temperature T_s.

$$t_g \le \frac{E_p}{W_m} + t_p$$

Substituting the values of W_m and E_p from Equations (4.20) and (4.21) respectively, we get

$$t_{gap} \le \frac{\phi t_b - \frac{\psi}{r}(1 - e^{-r \times t_b})}{\phi - \psi e^{-r \times t_b}} + \frac{1}{r} ln \frac{T^H - p}{T_s - p} \qquad (4.22)$$

where, $t_p = \frac{1}{r} ln \frac{T^H - p}{T_s - p} = t_b$

$$t_{gap} \le \frac{\phi t_b - \frac{\psi}{r}(1 - e^{-r \times t_b})}{\phi - \psi e^{-r \times t_b}} + t_b \qquad (4.23)$$

Empirical Observations:
Unoccupied period from experimental data is calculated here, as an example. In this experiment $T^H = 28.5°C$ and let us consider the desired temperature $T_s = 23°C$. Therefore, the maximum unoccupied period t_{gap}, during which HVACs can be run, is obtained from Equation (4.23) by taking the values of characteristic constants from Table 4.1.

$$t_b = \frac{1}{4.59} ln \frac{28.5 - 16.7}{23.0 - 16.7} = 0.14 \text{ Hr.}$$

$$t_{gap} = \frac{8736 \, x0.14 - \frac{7194}{4.59}(1 - e^{-4.59 \, x0.14})}{8736 - 7194 e^{-4.59 \, x0.14}} + 0.14$$

$$t_{gap} = 0.097 + 0.14 = 0.237 \text{ Hr.} = 14.22 \text{ min.}$$

Table 4.2 Characteristic Constants of
Warming Profile obtained by Curve-Fitting of Experimental Data.

Warming of a Zone (AC OFF)		
p'	q'	r'
26.87	−5.596	0.1186

4.5.8.1 *Early Shut Off*

Due to thermal capacitance, it takes time for a zone temperature to change after an AC is switched off. It is also known that a thermostat controlled AC does not maintain the zone temperature exactly at the set temperature T_s, but usually within a band of $\pm 1°C$. Considering the average case of zone temperature at T_s, can energy be saved if we shut off AC little earlier than at the end of use of the space, such that the zone temperature does not rise beyond $T_s + 1°C$?

Let t_w denote the warming time from T_s to $T_s + 1°C$. The early shut-off by time t_e can be done without compromising thermal comfort, if $t_e \leq t_w$.

Estimation of Warm Up Time:
The thermal profile of the warming of a zone can be generated from experimental data and the equation for the warming time of a zone can be derived by using curve-fitting as discussed in Section 4.5.3.2.

$$t_w = \frac{1}{r'} ln \frac{T^L - p'}{T^H - p'} = \frac{1}{r'} ln \frac{T_s - p'}{(T_s + 1) - p'} \tag{4.24}$$

where, i) T^L (T_s in this case) is the temperature when the AC is switched off and the zone warmed up to T^H (($T_s + 1$) in this case) and ii) p', q' and r' are the characteristic constants as shown in Table 4.2 (note the similarity between Equations (4.12) and (4.24)).

For $T_s = 23°C$, the early shut-off time t_e can be calculated as

$$t_e \leq \frac{1}{0.1186} ln \frac{23 - 26.87}{24 - 26.87} = 1.9 \text{ min.}$$

The result is not surprising, as it involves warming up only by $1°C$. Therefore, we can conclude that not much benefit can be derived by switching of AC earlier than the end of use of the space without compromising thermal comfort.

4.6 Thermal Comfort Under Peak Demand Constraints

The need for reducing peak demand arises due to very high peak-demand tariff imposed by the utilities upon the consumer, if consumption exceeds the peak demand limit. Since flatter peak demand facilitates improved grid stability, users' participation in reducing peak demand is also a necessity. So, let us look into the techniques of maintaining thermal comfort under the constraint of peak demand limit. In addition, it is also important to investigate and estimate savings in energy by taking measures like i) reducing the set temperature within acceptable limits, ii) optimum pre-cooling, etc.

Peak power demand can be reduced by i) *shifting high wattage loads* to non-peak hours and by ii) *staggering their operation*, in case of cyclic loads. Both these interventions require scheduling of the loads without compromising the intended objectives.

Thermostat controlled electrical devices (TCED) like air-conditioners (AC), refrigerators and heaters do not operate continuously.

They follow a pattern of ON-OFF cycles to maintain the temperature within a band, usually $\pm 1°C$ around the set temperature or the set point. These TCE devices do not need human intervention for their continuous operation. As long as the desired thermal comfort is maintained, users will not be concerned about when they consume the energy required to perform their assigned functions. Recall the discussion in Section 4.2.2 that under uncoordinated individual thermostat control, it is possible that all these TCEDs run simultaneously during some time intervals. This can result in higher peak power demand. Said differently, coordinated scheduling of TCEDs can reduce the peak demand. But, it is to be kept in mind that while the goal of scheduling HVACs is reduction in peak demand, this is to be done without compromising thermal comfort. In other words, the scheduling algorithm should ensure that TCE devices maintain the space temperature within the desired thermal comfort-band $[T^L, T^U]$, under peak demand-power constraint.

Consider n zones in a commercial building or a home, where the temperature of each zone is maintained by one or more ACs. In this discussion we take AC as a representative TCED. A zone is considered to be a closed space like a room with a single AC or a large room assumed to be divided into thermally decoupled zones whose temperatures are maintained by one AC in each zone. Peak power constraint limits the amount of energy that can be consumed at a point of time. Therefore, in order not to violate the peak-demand constraint, only m out of n TCEDs can run at a time. The TCEDs are to be scheduled such that the temperature $T_i, i = 1, 2, \cdots, n$ in individual zones lies within the comfort-band $[T^L, T^U]$.

4.6.1   *Candidate Scheduling Policies and their Limitations*

The cyclic ON-OFF operating nature of TCEDs makes them suitable to be modeled as periodic tasks. Therefore, it is not surprising that existing literature [Vedova *et al.* (2010, 2011); Barker *et al.* (2012); Subramanian *et al.* (2012)] suggest modeling of TCED loads as real-time tasks and applying traditional scheduling policies like EDF (Earliest Deadline First) [Liu and Layland (1973)] and LSF (Least Slack-time First) [Liu (2000)]. In this model, the time duration for which a TCE device remains ON is equated with the execution time requirement C_i within a period P_i and P_i is the sum of C_i and L_i where, L_i is the time duration for which the device remains OFF. But, scheduling algorithms from the real-time domain are not suitable for scheduling TCED loads for the following features of these class of algorithms, which do not conform to the operation of most of the home/building appliances under discussion.

- A task is considered schedulable if it receives C_i unit of processor time every P_i, irrespective of *when* it receives C_i within P_i.
- They do not take into account the environmental parameter (temperature here) feedback when taking scheduling decisions.

In order to schedule thermostatically controlled electrical devices, allocating power for C_i units of time within P_i is not just enough. It is also important *when* this C_i is allocated within the timeline of P_i. It can be observed from Figure 4.9(b), that if the TCED is powered-ON continuously (i.e., the task is executed) for C_i units of time at the beginning of its period from t_0 to t_b, the zone temperature T_i remains within the comfort-band. But, if the task is preempted after it is executed till t_a and is powered-ON (equivalent to allocating CPU time) for the remaining execution time just at the end of the period $[t_i, t_0 + P_i]$, T_i goes above T^U, the upper limit of the comfort-band $[T^L, T^U]$. It is important to note that the reason behind this is the exponential nature of the thermal characteristics of TCEDs.

Thus, whereas traditional real-time scheduling algorithms consider the deadline D_i, maximum laxity L_i and duty cycle P_i as constants for a task, in the case of TCE devices these parameters are dynamic, because execution time of such devices depend on the existing temperature of the very environment controlled by them. From Figure 4.9(b), it can also be observed that for the same TCED task, the duty cycle should change dynamically from P_i to $(t_c - t_0)$ when preempted at $t = t_a$. Also, the maximum laxity L_i changes from $(t_0 + P_i - t_b)$ to $(t_c - t_a)$. Note that the duty-cycle of a TCED

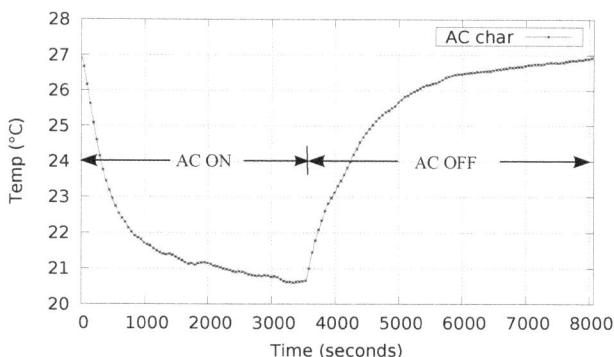

(a) Thermal characteristics of an AC

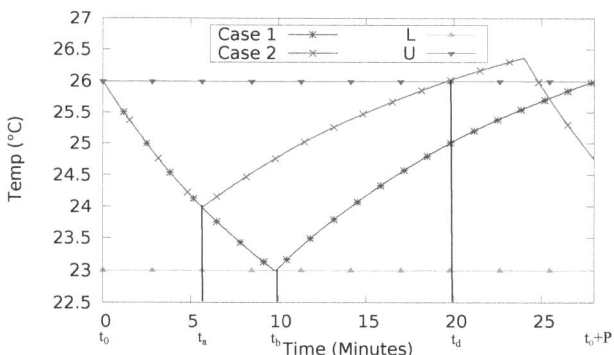

(b) TCED operation versus real-time task model

Fig. 4.9   TCED operation versus real-rime task model.

is related to the plant state variable (temperature). In case of real-time task there is no notion of duty-cycle, because real-time schedulability analysis deals with constant values of WCET (C_i) and P_i, and not concerned with plant state variable. Therefore, in order to apply any scheduling algorithm effectively for a set of TCED tasks, it is necessary that the parameters C_i and P_i are calculated dynamically. Furthermore, preemption decisions must consider the environmental parameter T dynamically, which is to be maintained within the desired comfort-band $[T^L, T^U]$. For example, so as not to violate the need for $T_i \leq T^U$ of a zone z_i maintained by AC A_i, another AC A_j, whose zone temperature T_j is $< T^U$ might have to be preempted to divert power to A_i.

Therefore, it is necessary to look for a scheduling algorithm, which is aware of the control variable, the space temperature. It requires an understanding of the thermal characteristics of the space controlled by TCE device like an AC.

4.6.2   *Thermal Characteristics of Heating, Ventilation and Air-Conditioning (HVAC) Systems*

Let us re-look at Figure 4.9(a) portraying the plot of temperature versus time generated from experimental data. It can be observed from Figure 4.9(a) that when an AC is on, i.e., when its compressor is running, it cools down the room temperature; the temperature rises from the instant the AC is switched OFF because of heat loads and losses. Observe that initially the cooling rate is very high and the cooling slope gets flattened exponentially as the temperature goes very low. For example, while it takes about 7 minutes to bring down the temperature from $27°C$ to $23°C$, the AC has to run for about 30 minutes to bring down the zone temperature from $23°C$ to $22°C$. This is due to the elementary principles of heat transfer that the rate of heat transfer is proportional to the difference of temperatures between two bodies (cooling coil of the AC and the air in contact with the coil surface, in this case).

It is known [ASHRAE (2001)] that i) finding characteristic constants (*thermal capacity* of the zone and *overall heat transfer coefficient*, etc.) is difficult, and ii) these constants vary widely with the changes in the ambient temperature, heat loads and losses. Therefore, online adaptation of thermal model of ACs, necessary for any scheduling/control mechanism to work effectively, is challenging. Due to these facts, the practical approach to generate AC thermal characteristics is driven by experimental data.

The thermal profile of an AC controlling the temperature of a given space can be generated experimentally by recording changes in the temperature of the space over time. From the elementary principles of heat transfer (discussed in Section 4.5.3.1) as well as from Figure 4.9(a) it can be stated that the thermal characteristic equation will take the form

$$T_i(t) = a + be^{-c*t} \tag{4.25}$$

where T_i denotes the temperature of ith zone corresponding to AC_i and a, b and c are the thermal characteristic constants of the AC for a given space with a given heat load and for a particular ambient temperature. The values of these thermal characteristic constants can be obtained by a suitable curve-fitting technique. Note that Equation (4.25) will be valid for both the cooling and the warming curves of an AC except for the values of

Fig. 4.10   Empirical data recording for AC characteristics (Sensor: RTD Pt 100 and Recorder: Eurotherm Chessell).

Table 4.3   Constants for AC characteristic equations (Ambient $= 27.5°C$).

(a) Cooling curve: AC ON			(b) Warming curve: AC OFF		
Constants	AC1	AC2	Constants	AC1	AC2
a	20.88	20.49	a'	26.87	27.43
b	6.174	6.357	b'	−6.17	−6.237
c	0.02092	0.01336	c'	0.01118	0.01141

the sign of the constants. In order to avoid confusion, let us denote that characteristic constants for warming curve as a', b' and c'. Table 4.3 shows an example of the thermal characteristic constants of two ACs obtained by curve-fitting, utilizing the measured temperature of two zones controlled by AC1 and AC2. The temperatures were recorded every 10 sec. in a digital recorder (Eurotherm Chessell 5000, shown in Figure 4.10) using $Pt100$ RTD (Resistance Temperature Detectors). These data will be used in the discussion on demand-response (D-R) control later in this chapter.

4.6.2.1   Cooling Down (AC Switched ON)

The cooling profile or the *cooling curve* of an AC can be represented by

$$T_i(t) = a + be^{-c*t} \tag{4.26}$$

Where, $T_i(t)$ is the temperature of the zone controlled by AC_i at time t and a, b, c are constants specific to particular ACs as shown in Table 4.3(a).

Solving for t, we get

$$t(T_i) = -\frac{1}{c} ln \frac{T_i - a}{b}$$

Therefore, the time for T_i to reach from T^U to T^L (equivalent to the execution time C_i of the task)

$$C_i = t(T^U) - t(T^L) = \frac{1}{c}\left(ln \frac{T^L - a}{T^U - a} \right) \tag{4.27}$$

Now, we can obtain the cooling slope S_c^i of AC_i from Equation (4.26) as

$$S_c^i = \frac{dT_i}{dt} = -bce^{-c*t} \tag{4.28}$$

$$\frac{1}{c}S_c^i = -be^{-ct} - a + a = a - (a + be^{-ct})$$

Considering the formulation of T_i (Equation (4.26)),

$$S_c^i(T_i) = c(a - T_i) \tag{4.29}$$

Equation (4.29) offers a simple approach to find the cooling slope at any temperature T_i. The cooling slope S_c at any given temperature T is a useful parameter as it is an indicator of cooling efficiency. The higher the slope, the lower the time an AC runs to bring down the temperature of a space by one unit of temperature. It may be noted that the value of a is equal to the lowest value of the zone temperature that can be achieved by running the AC.

4.6.2.2 Warming Up (AC Switched OFF)

The warming profile or the *warming curve* of an AC can be represented by

$$T_i(t) = a' + b'e^{-c'*t} \tag{4.30}$$

where, $T_i(t)$ is the temperature of the zone controlled by AC_i at time t and a', b', c' are constants specific to a particular AC as shown in Table 4.3(b).
From Equation (4.30), we get

$$t(T_i) = -\frac{1}{c'} ln \frac{T_i - a'}{b'}$$

Following the similar approach as followed in Section 4.6.2.1 we get the warming slope S_w^i as follows.

$$S_w^i(T_i) = c'(a' - T_i) \tag{4.31}$$

Equation (4.31) gives us a simple method to find the warming slope at any temperature T_i. It may be noted that the value of a' is the highest value of the temperature of the zone that can be reached, if the AC is not running.

It is to be noted that the following conditions always hold because of the exponential nature of the cooling and warming curves as it can be observed from Equations (4.29) and (4.31) as well as from Figure 4.9(a).

$$\forall\ \delta \geq 0,\ S_c^i(T_i) \leq S_c^i(T_i + \delta) \tag{4.32}$$

$$\forall\ \delta \geq 0,\ S_w^i(T_i) \geq S_w^i(T_i + \delta) \tag{4.33}$$

The above observations related to cooling and warming slopes are key to the formulation of the Thermal Comfort Band Maintenance (TCBM) algorithm for scheduling thermostat controlled electrical devices (TCED), as discussed in the following section.

4.6.3 *Analysis of Feasibility — Maintaining Thermal Comfort with TCBM Scheduling*

We will now discuss the maintenance of thermal comfort in n zones or spaces by n independent thermostat controlled ACs in each zone. With a peak demand power, limit power is available only for $m(< n)$ ACs at any point of time.

4.6.3.1 *TCBM Scheduling*

As the name suggests, the main goal of TCBM is to maintain the temperature of a zone within a desired band $[T^L, T^U]$, termed as *comfort band*. TCBM turns ON ACs judiciously so that not more than m out n ACs are running at any point of time in order to meet peak power constraint.

In order to maintain the comfort band, TCBM allows an AC to remain ON till its zone temperature reaches T^L or it becomes unavoidable to switch it OFF in the event of the temperature of some other zone reaching T^U. An AC_i, once switched OFF, is not turned ON till T_i reaches T^U. Before switching ON an AC whose zone temperature has reached T^U, another AC may have to be switched-OFF so that not more than m out of n ACs run at any point of time. This is necessary to meet the peak power demand constraint. Further, whenever it becomes necessary to switch OFF an AC to make power available to some other AC, it is logical to switch OFF the *coolest* available AC as defined below.

Coolest AC is the one among the running ACs, whose zone temperature will take maximum time to reach T^U, if switched OFF. Preference is given to the AC having lowest zone temperature, if more than one ACs meet this condition of $T_i < T^U$.

Note that it is desirable to keep the switching of ACs to as minimum as possible. This is because, every switching-ON of a compressor (an induction machine) in AC/refrigerator involves high starting current leading to additional power consumption. Further, frequent switching can affect the health of the equipment. In addition, a real-world equipment like AC compressor cannot be switched on and off at a high frequency like a real-time task. A running AC, once switched-off, can only be restarted after a minimum delay, usually 3 secs. Switching OFF the coolest AC will allow it to remain OFF for a longer time before its zone temperature reaches T^U and thus make it possible to achieve smaller number of switching in a given period of time.

TCBM Algorithm

Space cooling using AC involves bringing down the temperature to its desired value and maintaining it within some tolerance limits, usually $\pm 1°C$ of the set temperature. TCBM scheduling takes care of both i) *Cooling down* phase and ii) *Comfort-Band (CB) maintenance* phase.

- *Cooling down* phase: This is the phase when any zone temperature T_i is above the upper limit of CB ($T_i > T^U$) and TCBM scheduling of ACs brings down the temperatures of all the zones within the CB.
- *CB maintenance* phase: Once within the CB, the temperatures of all n zones are maintained in this phase by scheduling ACs using TCBM.

At the initial stage, when T_i is almost equal to T_a, an AC is switched ON to bring down the T_i from T_a to a value that lies within the comfort-band $[T^L, T^U]$. Specifically, the TCBM algorithm starts its *cooling down* phase by switching ON any $m(\leq n)$ numbers of ACs from the given set of n ACs. Considering the fact that wattage (the power demand) of the ACs can vary, the worst-case value of m is determined by the peak demand constraint of K watts.

$$\sum_{i=1}^{m} W_i \leq K \qquad (4.34)$$

where, W_i denotes the wattage of the ith AC and the sum is obtained from the first m of the n ACs arranged in descending order of their individual power requirement.

In order to maintain the thermal comfort band $[T^L, T^U]$, the following rules are applied by TCBM for scheduling ACs.

(1) *Rule # 1:* Turn OFF AC_i if it is ON at time t and if

   (a) $T_i \leq T^L$ *OR*

   (b) there is an $AC_j (i \neq j)$ with $T_j \geq T^U$ *AND* no. of ON-ACs $\geq m$ *AND* $T_i < T^U$ *AND* AC_i is the *coolest* one among ON-ACs.

(2) *Rule # 2:* Turn ON AC_i if

   (a) $T_i \geq T^U$ *AND*

   (b) No. of ON-ACs $< m$

Rule # 1(a) ensures that AC_i is switched OFF when the zone temperature T_i reaches its lower limit T^L. This is because, running an AC below T^L not only over-cools affecting the thermal comfort, but also leads to wastage of energy. However, the need may arise to switch OFF an AC even before its zone temperature reaches T^L, in case it is observed that the zone temperature of some other AC reached T^U. This is necessary to provide power to the AC, whose zone temperature reached T^U and thereby enabling it to maintain the desired thermal comfort. Rule # 1(b) takes care of such a scenario. Rule # 2 ensures that if a zone temperature reaches its upper limit T^U, the AC belonging to the zone is switched ON subject to the condition that not more than $m(\leq n)$ ACs run at any point of time.

4.6.3.2 *TCBM Feasibility for Cooling*

The idea behind TCBM feasibility is simple. It ensures that the aggregate cooling by m ACs are more than the aggregate warming due to $(n - m)$ ACs, which are not running [Karmakar *et al.* (2015a)].

If peak-demand constraint allows m out n ACs to run at a time and ACs are scheduled by the TCBM algorithm, all the zone temperatures will eventually fall from the ambient T_a to a value T_i ($T^L \leq T_i < T^U$) and the CB $[T^L, T^U]$ will be maintained thereafter, if the following condition is satisfied.

$$\sum_{i=1}^{m} abs(S_c^i(T_i)) > \sum_{i=1}^{n-m} S_w^i(T_i) \qquad (4.35)$$

where, i) the sum on the left is obtained from the first m of the n ACs arranged in ascending order of their cooling-slopes, ii) the sum on the right is obtained from the first $(n-m)$ of the n ACs arranged in descending order of their warming-slopes and iii) $m \geq \lceil n/2 \rceil$.

Note that absolute value of $S_c^i(T_i)$ is taken because the cooling slope is negative.

4.6.3.3   TCBM Feasibility for Heating

A TCE heating load (e.g., room-heater) functions exactly in the opposite manner compared to a TCED cooling load like an AC. A heater must be switched ON whenever the zone temperature T_i goes below $T^L (> T_a)$. A heater must be switched OFF, if $T_i \geq T^U$. Here the assumption is that each zone is heated by a single heater. Therefore, for feasibility of running m out of n TCE heaters under peak demand constraint, it is to be ensured that the aggregate heating by m heaters are more than the aggregate cooling due to $(n-m)$ heaters which remain off at any point of time.

If peak-demand constraint allows m out n heaters to run at a time and the heaters are scheduled under TCBM algorithm, all the zonal temperatures will eventually rise from the ambient T_a to a value T_i $(T^L \leq T_i < T^U)$ and the CB $[T^L, T^U]$ will be maintained thereafter, if the following condition is satisfied.

$$\sum_{i=1}^{m} S_w^i(T_i) > \sum_{i=1}^{n-m} abs(S_c^i(T_i)) \tag{4.36}$$

where, i) the sum on the left is obtained from the first m of the n heaters arranged in ascending order of their heating-slopes, ii) the sum on the right is obtained from the first $(n-m)$ of the n heaters arranged in descending order of their cooling-slopes and iii) $m \geq \lceil n/2 \rceil$.

4.6.4   Analysis of Energy Consumption

The energy consumption of an AC not only depends on the numerical value of the $^\circ C$, by which it brings down the zone temperature, but also on the initial (T_1) and the final (T_2) temperatures. This is because, the ON-time $(t_2 - t_1)$ of an AC can vary even if the amount of change in temperature $(\Delta T = T_1 - T_2)$ caused by it remains the same. For example, it can be observed from Figure 4.9(a) that time required by the AC to bring down the temperature from $26^\circ C$ to $24^\circ C$ is much lesser than the time required for cooling the zone from $24^\circ C$ to $22^\circ C$. This happens because of the exponential nature of the cooling curve having a variable slope.

The cooling slope S_c is a measure of the drop in temperature T per unit time. From Equation (4.29), the cooling slope at any temperature T can be derived as

$$S_c = c(a - T) \qquad (4.37)$$

It can be observed from Equation (4.37), that the slope (S_c) of the cooling curve is a function of the zone temperature T and S_c reduces linearly with the fall in zone temperature. The plot of temperature versus cooling slope corroborates this fact as shown in Figure 4.11(a).

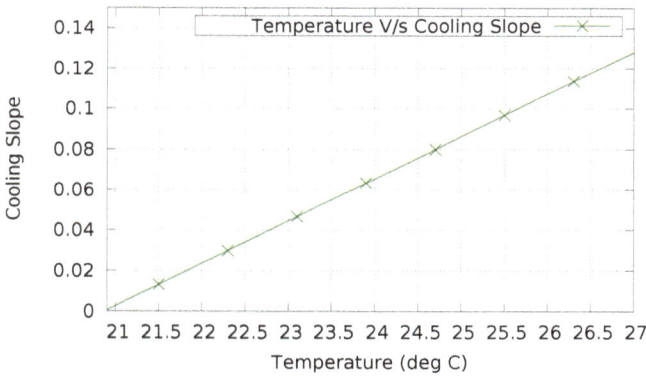

(a) Cooling slope varies with zone temperature

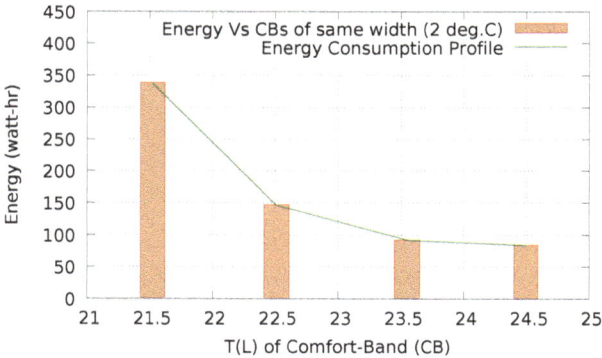

(b) Energy consumption with variation in CB

Fig. 4.11   Change in cooling slope and energy consumption with variation in zone temperature. [Karmakar *et al.* (2015a)]

Also, from Equation (4.28),

$$\frac{S_c}{-bc} = e^{-c*t}$$

$$-ct = ln\frac{S_c}{-bc}$$

[Note that the RHS of the above expression is always a logarithm of a positive number owing to the fact that the cooling slope S_c is -ve, b and c are both positive (for a cooling curve).]

$$t = \frac{1}{-c}ln\frac{S_c}{-bc} \qquad (4.38)$$

The energy consumption (E) of an AC running for a time interval of $[t_1, t_2]$ can be expressed as

$$E = \int_{t_1}^{t_2} W\,dt$$

where, W is the wattage of an AC.

This leads to

$$E = W[t_2 - t_1]$$

Substituting t from Equation (4.38),

$$E = W \times \frac{1}{-c}\left(ln\frac{S_c(t_2)}{-bc} - ln\frac{S_c(t_1)}{-bc}\right) \qquad (4.39)$$

$$E = \frac{W}{c}ln\frac{S_c(t_1)}{S_c(t_2)}, \qquad (4.40)$$

let us assume that at t_1, when the zone temperature is T^U, the AC is switched ON, and at t_2, when the zone temperature is T^L, the AC is switched OFF.

This allows us to rewrite Equation (4.40) as

$$E = \frac{W}{c}ln\frac{S_c(T^U)}{S_c(T^L)} \qquad (4.41)$$

Equation (4.41) leads us to conclude that as the ratio $\frac{T^U}{T^L}$ increases, the energy consumption increases exponentially even if the comfort-bandwidth $(T^U - T^L)$ remains the same.

The computed energy consumption data pertaining to cooling down of temperature from T^U to T^L for different comfort-bands of equal width $[(T^U - T^L) = 2°C]$ are presented in Figure 4.11(b).

From Figure 4.11(b), it can be observed that the energy consumption can be reduced significantly by shifting the comfort-band up just by $1°C$ e.g., from $[23.5°C, 21.5°C]$ to $[24.5°C, 22.5°C]$.

4.6.4.1 *A Case Study on Energy Savings*

The results of a real world implementation are presented here demonstrating energy-saving by shifting of comfort-band. The set up is shown in Figure 4.12. TCBM algorithm is applied on 2 ACs installed in a room and experimented for two different comfort-bands. It is assumed that a peak demand limit puts a constraint, which allows only 1 AC to run at a time. The TCBM analysis applied on the thermal characteristics of the ACs established that in order to maintain the comfort-bands, it is feasible to run 1 out of 2 ACs at a time. It may be noted that thermal characteristics of the ACs were obtained while the ambient temperature was $27°C$.

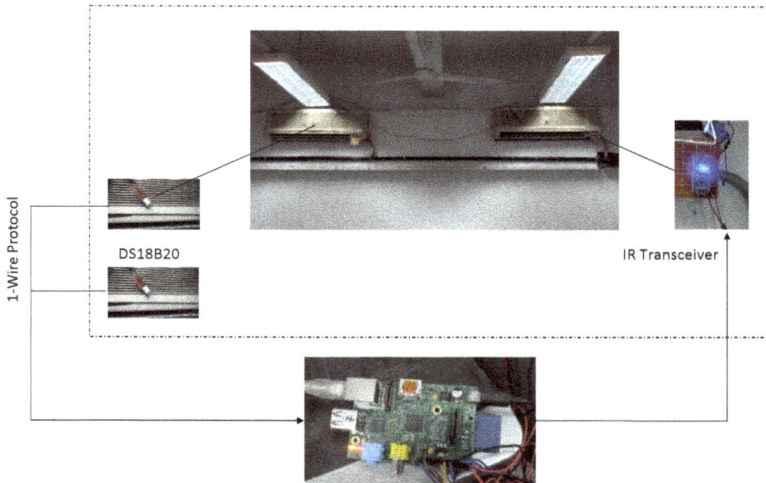

Fig. 4.12   TCBM scheduling implemented in Raspberry Pi [Equipment within dotted rectangle are the sensors, actuators and the controlled equipment (TCEDs)]. [Karmakar *et al.* (2015a)]

The experiment was started with a CB $[24°C, 22°C]$ and then it was shifted up by $0.5°C$ to a CB $[24.5°C, 22.5°C]$. It is assumed that a small change in temperature $(0.5°C)$ will not affect the thermal comfort of the occupants. The duration of the experiment was 1 hour in both the cases.

Figures 4.13(a) and 4.13(b) show the operation of ACs under TCBM control for CB $[24°C, 22°C]$ and CB $[24.5°C, 22.5°C]$ respectively. The cumulative on-time (COT) of the ACs are computed for both the cases. As it can be observed from Figure 4.13 that shifting the CB by $0.5°C$ led to 11 minutes of reduction in COT.

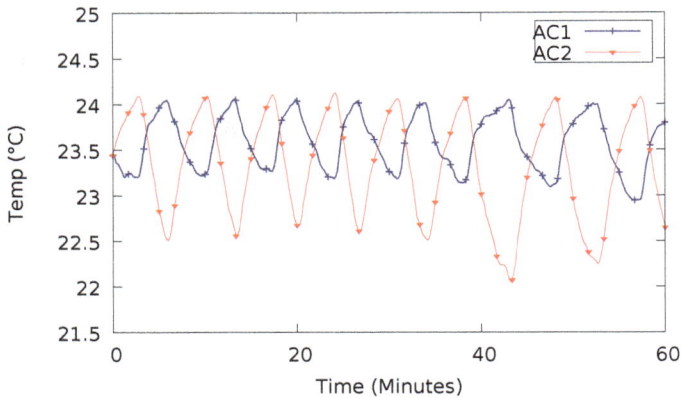

(a) CB 24°C - 22°C with COT = 60 *min.*

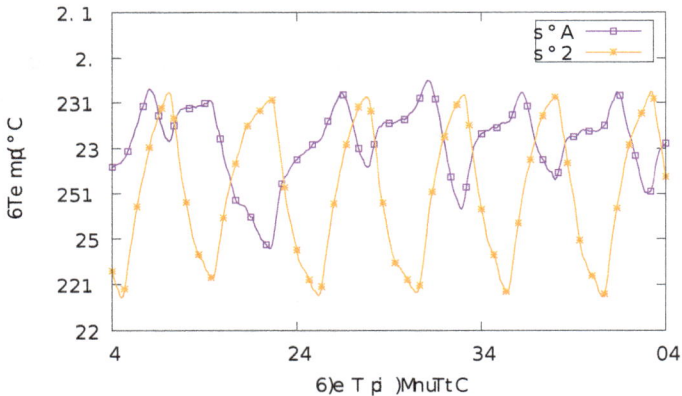

(b) CB 24.5°C - 22.5°C with COT = 49 *min.*

Fig. 4.13   Effect of CB on energy consumption: TCBM scheduling of 2 ACs for different CBs. [Karmakar *et al.* (2015a)]

Energy consumption is directly related to the cumulative ON-time (COT) of the ACs. The results presented in Figure 4.13 show about 18% saving in an hour in the small prototype system of 2 ACs. Considering the energy consumption of a 1.5 ton AC as 1.8 KW and assuming that the room is occupied 10 hours/day on an average, the saving in energy is

$$\frac{11}{60} \times 10 \times 30 \times 1.8 = 99 \text{ units (KWh) per month.}$$

4.7   Adaptive Demand-Response (D-R) Control

Balance between generation of power and its consumption is the key to the stability of the electrical power system. Utilities prefer a flatter demand curve throughout the day as it helps them in reducing its reserve generation capacity to meet the peak demand in certain times of a day. Therefore, non-uniform electricity tariffs are imposed upon the consumers, i) to recover the additional investment towards meeting peak demand and ii) to influence customers' consumption pattern. In this section, various approaches in D-R control are discussed.

4.7.1   *Adapting Energy Consumption with TOD Charges*

Supply utilities bill the consumers typically under two headings — energy charges and demand charges. In many countries, TOD (time-of-day) and peak-demand electricity charges [Power (2014); Maharashtra State Electricity Distribution Company, India (2014)] are applicable to bulk consumers, which include commercial buildings and academic institutions. However, in developed countries, small residential consumers also come under TOD and peak demand tariffs. It is likely that, in the near future, small residential consumers in the developing countries will also come under similar tariff policy. Further, peak demand charge may vary depending on the time of a day.

Therefore, consumers who want to reduce their electricity bill, need to adapt to the peak power demand limit by limiting energy consumption demand during some period of the day due to TOD charges.

Changes in the ambient temperature and occupancy affects the energy consumption of the TCEDs as it alters the thermal characteristics (represented by the characteristic constant discussed in Section 4.6) of these devices. Further, the experimental results in Section 4.6.4.1 establishes that even a small shift in the CB by $0.5°C$ can cause a significant change in energy consumption.

Therefore, techniques for controlling TCE devices are required for adapting to i) changes in the thermal characteristics of an AC under different ambient conditions and ii) maintaining the desired comfort-band under time varying peak power limit.

Note that changes in thermal characteristics results in changes in the cooling (S_c^i) and the warming (S_w^i) slopes. On the other hand, changes in peak demand affect the value of m, the maximum number of TCEDs that can run at a time (Equation (4.34)). Thus, the schedulability of TCEDs gets affected under both these cases (Equation (4.35)).

A discussion in Section 4.6.4 shows how energy consumption can vary with shifting of the comfort-band. Therefore, the electricity bill can be reduced by adjusting the comfort-band according to the TOD charges. For example, comfort-band can be shifted up slightly, say by $0.5°C$ during 9 am to 12 noon, when TOD charge is higher.

4.7.2   *Handling Varying Ambient Temperature and Occupancy*

Let us discuss i) a case study on the effect of varying ambient temperature on schedulability under peak demand constraint and ii) an adaptive demand-response (D-R) technique to handle it. Note that the same technique is equally effective in handling varying occupancy as well. This is because the effect of variation in ambient temperature and occupancy both get reflected in the form of changes in the slopes of cooling curve S_c and the warming curve S_w of an AC.

A Case Study: The effect of ambient temperature on the schedulability of m out of n ACs in maintaining a given comfort-band is studied. In is assumed that the peak demand limit is constant, but ambient temperature and occupancy can vary with time.

The zone temperature data of ACs on two different days were recorded when the ambient temperatures were $27°C$ and $30°C$, respectively. The thermal profile of an AC is characterized by its cooling slope S_c and the warming slope S_w. Table 4.4 shows the effect of changed thermal characteristics of ACs on schedulability. The schedulability data for 5 ACs shown in

Table 4.4   Schedulability with varying ambient (5 ACs).

Ambient $= 27°C$				
CB $(°C)$	m	$\sum S_c$	$\sum S_w$	feasibility $(\sum S_c \geq \sum S_w)$
$23 - 25$	2	0.149	0.140	Yes
$24 - 26$	2	0.218	0.098	Yes
Ambient $= 32°C$				
CB $(°C)$	m	$\sum S_c$	$\sum S_w$	feasibility $(\sum S_c \geq \sum S_w)$
$23 - 25$	2	0.35	0.53	No
$24 - 26$	2	0.46	0.35	Yes

Table 4.4 are generated by the feasibility criterion as in Equation (4.35). It can be observed from Table 4.4 that by running $2(= m)$ out of $5(= n)$ ACs at a time, it is possible to maintain a CB $[23°C, 25°C]$, when the ambient temperature is $27°C$. However, it is not so when the ambient temperature goes up to $32°C$. But, as it can be observed from Table 4.4, we have the following option of *adjusting the comfort-band to CB [24°C, 26°C] so that the peak demand limit of available power for 2 ACs is not violated.*

4.7.2.1   *Scheme to adapt to the changes in ambient temperature*

• *Scheme I*

  – Generate AC characteristics constants at various ambient temperature values obtained by off-line curve-fitting. (Here it is assumed that change in heat load is negligible.)
  – Store the data in TCBM controller and use it for adaptive control.

• *Scheme II*

  – Carry out on-line curve-fitting based on the room temperature obtained in the immediate past say, data of last 10 minutes obtained at regular intervals. It is assumed that there will be no major variation in ambient temperature within 10–15 minutes.
  – Use the data for feasibility analysis and apply adaptive control accordingly.

It is worth noting that Scheme II is also capable of handling variations in heat loads due to changes in occupancy as well as any addition or removal of heat producing equipments like computers, etc.

4.7.3   TCBM as Anytime Algorithm to Handle Varying Peak Limit

Algorithms whose quality of results vary with computation time are called *anytime algorithms*. The quality of results gradually improves with the increase in allotted computation time. The concept of imprecise computation was introduced in [Liu and Wei-Kuan (1995)] and the authors applied it to real-time systems. It is shown that the imprecise computation techniques offer scheduling flexibility with a compromise on the quality of the result. But, it helps meeting the computational deadline.

Given a peak demand, TCBM can either

i) compute schedulability and come out with a YES/NO answer for maintaining the desired comfort-band $[T^L, T^U]$, or
ii) it can compute what comfort-band can be maintained under the given peak demand limit.

Because of this flexibility, TCBM can be utilized as an anytime algorithm. If TCBM schedulability analysis concludes that the desired comfort-band cannot be maintained, it can offer a maintainable but slightly *inferior* thermal comfort-band than what is most desired by the consumer. We assume that the consumer may accept a slightly inferior comfort-band, as it involves financial incentive if power consumption is maintained within the peak limit.

4.7.3.1 *Imprecise Computation and Inferior Thermal Comfort*

An analogy between imprecise computation and variation in desired thermal comfort level, termed as *inferior thermal comfort*, is shown in Table 4.5.

Table 4.5   Analogy between Anytime Algorithm and TCBM Algorithm.

	Constraint	**Computational Output**
TCBM Algorithm	Peak Demand Limit	Inferior comfort band
Anytime Algorithm	Deadline	Imprecise

Consider a system of n ACs controlling n zones in a building. Given a peak demand constraint that allows only $m(\leq n)$ ACs to run at a time, TCBM schedulability analysis can tell us if it is feasible to maintain a given thermal comfort-band $[T^L, T^U]$ or not. If not, TCBM can also compute the comfort-band that can be maintained under the given peak limit. For this purpose all we have to do is to check for feasibility by increasing T^L and T^U say, by $0.5°C$, in each iteration.

As shown in Table 4.6, individual thermal comfort preference can be expressed in terms of 7-point PMV scale defined by ASHRE in [ASHRAE (1992)]. Though the desired thermal comfort is defined as 0, we assume that a consumer may be ready to accept a thermal comfort level of $+1$ or -1, if financial incentive is offered. After all, it is only a marginal compromise as $+1$ and -1 represent the PMV values corresponding to *slightly warm* and *slightly cold* state, respectively.

Table 4.6 7-point ASHRE thermal scale.

Vote	Thermal Comfort Level
+3	Hot
+2	Warm
+1	Slightly Warm
+0	Neutral
−1	Slightly Cold
−2	Cool
−3	Cold

Let us demonstrate the effect of shifting of CB on the TCBM feasibility (under constant ambient temperature) with the help of an example. Suppose, peak demand constraint allows power for 3 ACs at a time and we have 5 ACs, which can be controlled using TCBM algorithm. Let us consider that out of 5 ACs, two ACs have the same thermal characteristics as that of AC1 and three ACs have the same characteristics as that of AC2. The constants of their characteristic equations are shown in Tables 4.3(a) and 4.3(b). The data related to feasibility of these 5 ACs under TCBM are generated using Equations (4.29) and (4.31) as shown in Table 4.7. It can be observed from Table 4.7, that it is feasible to maintain a comfort-band of $[23°C - 25°C]$ by running 3 ACs at a time. But, the same comfort-band cannot be maintained, if peak demand constraint allows powering-ON of at most 2 ACs at a time. Now, let us shift the comfort-band from $[23°C, 25°C]$ to $[24°C, 26°C]$. It can be observed from Table 4.7 that in this case, the comfort-band can be maintained by running 2 out of 5 ACs at a time. This leads to the conclusion that under varying peak demand limit, the comfort-band $[T^L, T^U]$ can be suitably shifted to meet the peak demand constraint.

Table 4.7 Schedulability by shifting comfort-band (5 ACs).

Comfort-band $(°C)$	m	$\sum S_c$	$\sum S_w$	feasibility $(\sum S_c \geq \sum S_w)$
23 − 25	3	0.72	0.34	Yes
23 − 25	2	0.35	0.53	No
24 − 26	2	0.46	0.35	Yes

A TCBM-based *inferior thermal comfort* algorithm that has the potential for users' participation in demand-response (D-R) control is shown in Figure 4.14.

$n = $ *Total no of* ACs *and* $m = $ *Maximum number of* ACs *that can run at a time*

$PMV = 0$

$T^S = 24;$ *Default Temperature Set Point* $(^\circ C)$

Comfort-Band (CB) $[T^L, T^U] = [T^S - 1, T^S + 1];$ *Default Comfort-Band*

BEGIN

$t = 0, \quad t_1 = 0$

DO

read $t,$ PMV *and calculate* m *from given peak demand constraint* W^P

IF $((t - t_1) \leq 10)$ *minutes* **CONTINUE**; **ELSE** $t = t_1;$ **END IF**; *Checking every* 10 *minutes*

IF *(PMV > 0)*   **THEN**

$\qquad T^L = T^L - 0.5$ *and* $T^U = T^U - 0.5$

$\qquad$ *run* TCBM *feasibility analysis for* m *and* $CB[T^L, T^U]$

$\qquad$ **IF** *feasible,* **CONTINUE**;

$\qquad$ **ELSE**

$\qquad\qquad$ *notify Consumer*

$\qquad\qquad$ **IF** *customer agrees to accept slightly less (inferior) comfort* **THEN**

$\qquad\qquad\qquad T^L = T^L + 0.5$ *and* $T^U = T^U + 0.5;$

$\qquad\qquad$ **END IF**

$\qquad$ **END IF**

ELSE IF *(PMV < 0)*   **THEN**

$T^L = T^L + 0.5; \quad T^U = T^U + 0.5;$

END IF

WHILE(1)

END

Fig. 4.14   TCBM inferior comfort algorithm.

4.7.4   *Adaptive Demand-Response Policy*

The discussion above leads to the following DR (demand-response) policy.

(1) Adjust the comfort-band dynamically and inform the user about it on the event of the following:

- Changes in the externally-imposed peak demand constraint,
- Changes in the ambient temperature, and
- Changeover to the TOD slots with different charges imposed by the distribution company.

(2) If a user insists on staying with a pre-set comfort-band, he can be warned about the implications of violating the peak power consumption limit or consuming more during higher TOD-charge slot ahead of time.

4.8 Learnings from an Academic Building

The holistic approach of providing thermal comfort in building spaces discussed earlier in this chapter is applied to four different types of buildings spaces as shown in Table 4.8. We will now discuss the learnings from this exercise.

Table 4.8   Description of spaces under case study.

| Space | Architecture | | Design |
	Size	Floor	Conditioning Resources
Auditorium	Large	Slant	HVAC
Big Classroom	Large	Flat	HVAC, Fans, Windows
Small Classroom	Small	Flat	HVAC, Fans
Lab	Medium	Flat	HVAC, Fans, Windows

4.8.1 *Auditorium*

The auditorium under consideration is a large space with slanted floor and maximum occupancy of 200 people. For this enclosed space 6 (six) AC units attached to two air mixing plenums are provisioned for climate control. The conditioned air is passed through the octagonal shaped chilled beam into the space via eighteen ducts, out of which four are on the inner side, closer to middle area, whereas the remaining are on the outer side as illustrated in Figure 4.15. All these ACs are traditional thermostat controlled ON-OFF type devices. Note that such ACs are going to exist for some more years in existing buildings, until they are replaced by more efficient VRF (variable refrigerant flow) type of ACs, due to the feasibility of retrofitting and economic reasons. However, the discussions in this section are general and independent of the types of ACs deployed for thermal conditioning of a space.

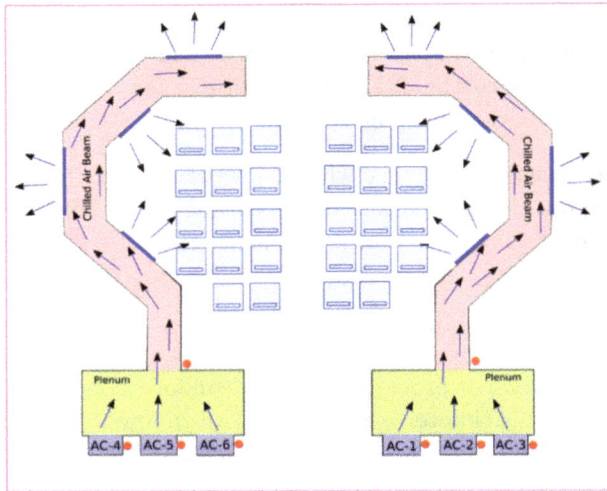

Fig. 4.15   Illustration of auditorium.

4.8.1.1   *Thermal Properties of the Space*

The deployed HVAC system affects the auditorium thermally depending upon how many AC units are operational and for how long. The *Thermal Gradients* and the *differences in the Cooling times* are observed as discussed below.

Thermal Gradient: Figure 4.16 shows the variations in temperature in the subspaces within the auditorium — faster cooling in the area around the middle row (*row5*) compared to the rows on either side, front (near the podium) and back (near the exit) due to the structure of the beams and the duct openings. The attained temperature is usually 1°C–2°C lower than the temperature in other areas. The blue lines correspond to cooling curve and red lines indicate warming curve, i.e., the ON and OFF periods of the duty cycle of an AC respectively.

Differences in Cooling Time: The following observations can be made from Figure 4.17.

- Reduced variation in the temperature profile of rows in the auditorium as the number of AC units in operation increases. The area of cooling also varies, as seen in the figure — with four AC units only *row5* cools faster, whereas with five units additional rows cool faster.

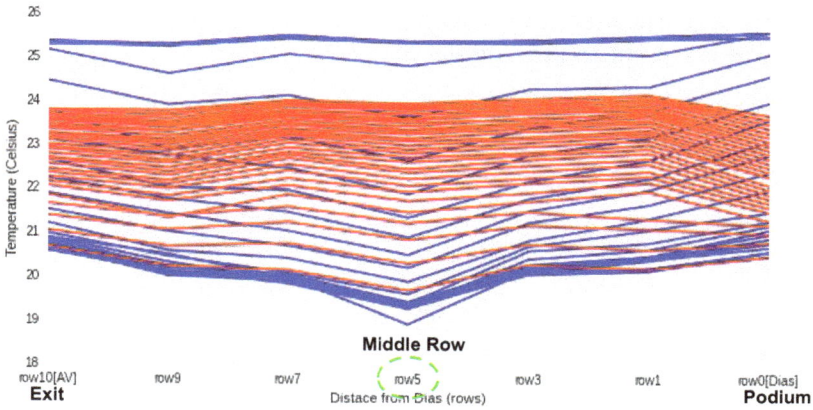

Fig. 4.16 Temperature gradient observed in auditorium with four ACs.

- *row0[Dias]* cools later than other rows irrespective of the number of ACs used.
- The time required for cooling *row5* is more with four ACs than with six ACs.

4.8.1.2 *Interventions and their Benefits*

The above observations indicate the need for interventions in order to achieve the necessary conditioning of the space depending on where the occupants are seated.

- *Seating Arrangement Policy*: With thermal gradient data, we come to know where the cooler and warmer subspaces exist in the space. Seating people in the cooler areas first will improve their state of comfort. For this auditorium, since its middle area cools faster, arranging people around the middle row (*row5*) is the suggested (energy conscious) seating policy in case of partial occupancy.
- *Number of AC Units*: From Figure 4.17, it is possible to know which *rows* cool faster when a certain number of AC units are used. Assuming that people follow the prescribed energy conscious seating policy, the controller can act intelligently to decide how many AC units are to be used during an event.
- *Pre-Cooling Time*: In case of full occupancy, the row wise temperature profiles help to pre-compute cooling times [Karmakar *et al.* (2015a)].

This helps the controller to know precisely when to turn on AC units before the event to pre-cool the space.

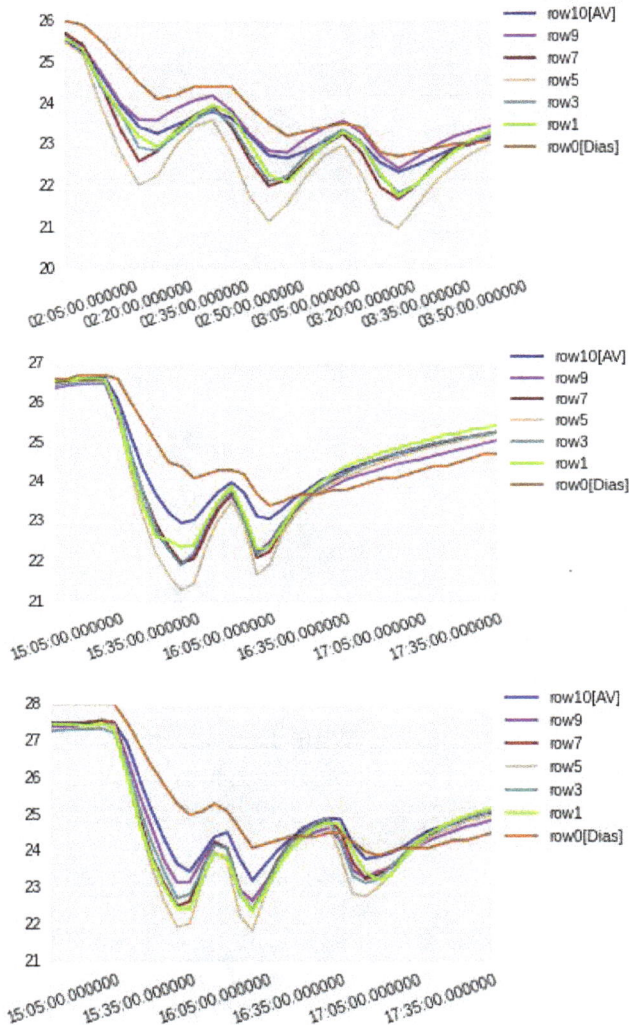

Fig. 4.17   Impact of varying the number of AC units: with 4 ACs (top), 5 ACs (middle), and 6 ACs (bottom).

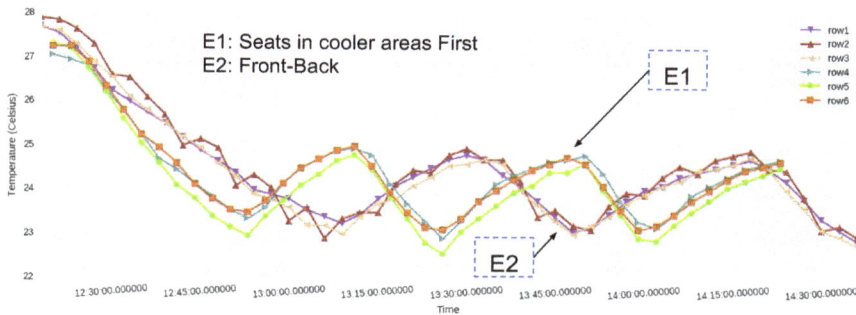

Fig. 4.18   Comparison of temperature profiles between energy conscious (E1) and front-back (E2) seating arrangement.

Benefits

Benefits of these interventions can be observed by comparing the suggested energy conscious seating policy with the usual front-to-back seating policy.

(1) Front-Back Arrangement: People occupy seats starting from first row to last row.
(2) Energy Conscious Seating Arrangement: People occupy seats from the middle row (*row5*) and spread in the outer direction on either side.

Figure 4.18 reveals the following information.

- Front-Back Seating Arrangement (*row1*, *row2*, *row3*) takes longer to attain 23°C compared to Energy Conscious Seating Arrangement (*row4*, *row5*, *row6*)
- The length of the ON period of the AC duty cycles, is longer than that of the Energy Conscious Seating Arrangement.

With people seated in front-back arrangement, they occupy *row1*, *row2* and *row3* (refer Figure 4.18). The maximum of cooling times of these *rows* is more than an hour to bring the temperature from 27°C to 23°C which can be noted from Table 4.9. Further, Thermal Comfort Band Maintenance (TCBM)[6] [Karmakar *et al.* (2015a)] approach is used to maintain the temperature within 25°C to 23°C band with average cycle (without pre-cool cycle) energy consumption of 8045 W-Hr.

---

[6]Turning OFF AC when lower threshold is reached, turn ON AC when upper threshold is reached, thus maintaining temperature within a band.

Table 4.9   Energy consumption for different seating arrangements in the auditorium — note that energy conscious seating reduces consumption by more than 30.

Seating Arrangement	Pre-cool Time (mins) (27°C to 23°C)	Avg. Cycle Time (mins)	Avg. Cycle Energy Consumption (W-Hr)
Front-Back	72	52	8095
Energy Conscious	49	37	5195

With energy conscious seating arrangement, *row4*, *row5* and *row6* (refer to Figure 4.18) are occupied with maximum cooling times of occupied rows less than an hour (refer Table 4.9). The average cycle (without pre-cool cycle) energy consumption is 5195 W-Hr which is 35% lesser compared to front-back seating arrangement. Table 4.9,[7] shows the pre-cool and average cooling-heating cycle time for each of the seating arrangements.

This case study demonstrates that the systematic approach discussed in this section can lead to better thermal conditioning of the space at a lower energy cost.

4.8.2   Small Classroom

The small classroom space, with a flat floor, is provided with HVACs and fans to control the indoor climate and is equipped to handle a class of maximum 40 people. The HVAC is a Variable Refrigerant Flow (VRF) based system with three indoor units (IDU) conditioning this space. Four ceiling fans are provided above the seating locations for air circulation.

4.8.2.1   Thermal Properties of the Space

The conditioning resources installed in a space influence its thermal property depending on which resource is ON, what temperature controller is in use, etc. With respect to this, the following observations are made

- *Over-Cooling*:
  The ACs tend to cool the space below the desired set point while IDUs are in operation as pointed out in Figures 4.19 and 4.20. The behavior is often seen when IDUs are turned ON after being OFF for the night.
- *Nonuniform Temperature Distribution*:
  When IDUs are operated without fans we see a clustering effect in

---

[7]These energy calculations were carried out considering a static occupancy as varying occupancy can affect energy consumption.

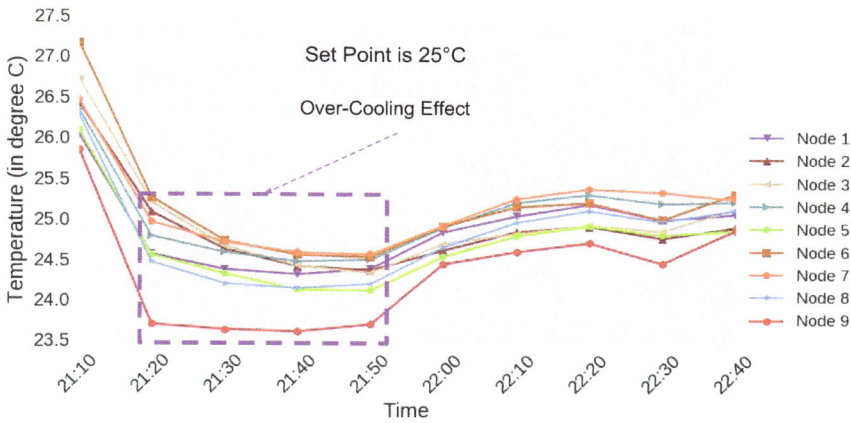

Fig. 4.19   Effect on thermal profiles in the absence of fan circulation.

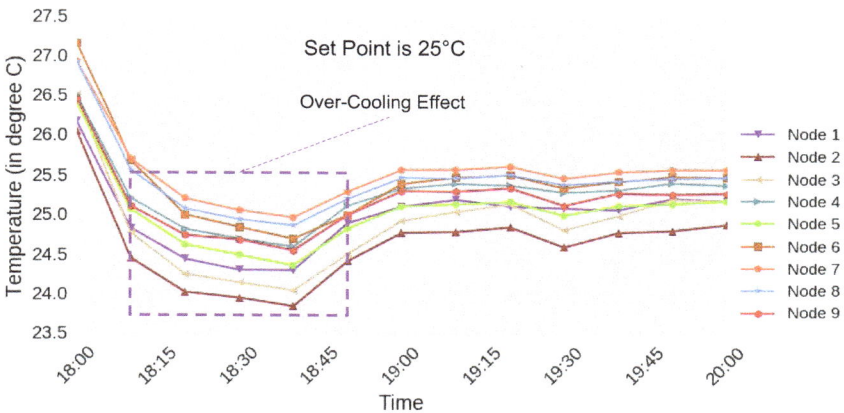

Fig. 4.20   Effect on thermal profiles in the presence of fan circulation.

thermal profiles of subspaces[8] in the classroom. Few subspaces cool faster whereas some do not, as seen from Figure 4.19.

---

[8]A network of temperature nodes was laid in the classroom to capture its thermal profile.

Figure 4.19 shows that

- Non-uniform distribution of temperature can be prominently high and also,
- The over-cooling action causes the room temperature to drop to 24°C though the set-point is 25°C

Figure 4.20 shows that

- The clustering effect which is seen in Figure 4.19 in the form of non-uniform temperature distribution is negated with the use of fans. The temperature sensor nodes show a uniform spread of temperature values, but,
- Over-cooling effect is seen in this case as well.

4.8.2.2   *Interventions and their Benefits*

To tackle the problem of over-cooling, a custom temperature control algorithm is employed. This custom algorithm maintains temperature around the desired set point, the effect of which is seen in Figure 4.21. As a result, over-cooling does not take place, also additional energy is not expended.

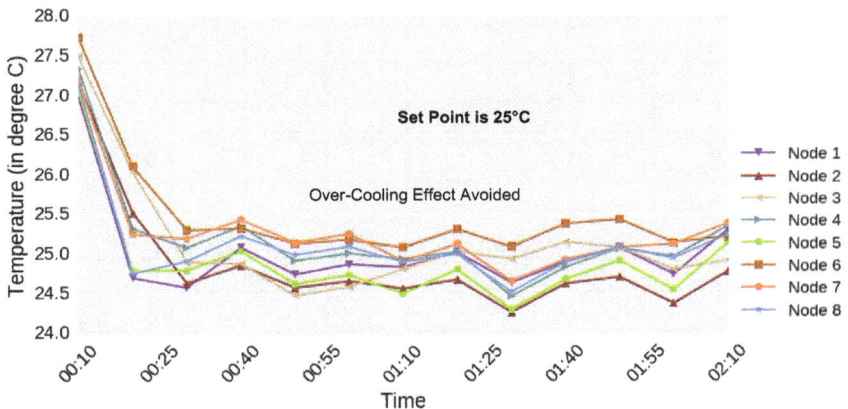

Fig. 4.21   Custom temperature controller negates over-cooling affect.

It can be observed from Figure 4.21 that

- Use of custom algorithm prevents over-cooling effect observed in Figures 4.19 and 4.20, and
- Also a uniform spread of temperature values is achieved.

4.8.3 *Big Classroom*

The big classroom is a large space with a flat floor surface and windows, unlike the auditorium. All conditioning resources, HVAC, fans and openable windows are deployed to control the indoor climate. Like the small classroom, HVAC is VRF based system with six IDU's with louvers and connected to three[9] compressor or outdoor units (ODU) along with eleven ceiling fans. This space is equal to the size of two small classrooms with maximum occupancy of 80 people and IDU's and ceiling fans divided among these two halves.

4.8.3.1 *Thermal Properties of the Space*

The combination of the conditioning resources provided in the space shows the following thermal observations depending on the location of IDU's, which IDU's are ON, etc.

- *Over-Cooling*:
  Since the HVAC system in this space is the same as the small classroom, same over-cooling effect is seen here as described in *Thermal Properties* of small classroom in Subsection 4.8.2.
- *Nonuniform Temperature Distribution*:
  Different IDU's cater to two different subspaces leading to the development of zones in the given space. IDU's have a distinct effect on these zones, leading to some zones cooling faster than the others.

4.8.3.2 *Interventions and their Benefits*

- Custom controller intervention described in Subsection 4.8.2 is used as is in the big classroom, over IDU's in different zones.
- Run Time control leveraging *Zone Based Cooling* and *One Zone At a Time Seating Policy* are mentioned in Algorithm 3.

---

[9]There are 18 more IDU's in five other rooms.

Algorithm 3 Intervention: Leveraging Zone Based Cooling

---

if Occupancy $\leq$ #Seats in Half of Big Classroom **then**
    Occupy the seats in first half
else
    Start occupying seats in first half, open next half only when first half
is full. Turn ON all IDUs and fans
end if

4.8.4 *Lab Space*

The lab is a medium-sized space provisioned with operable windows and
exhaust fans along with two IDU's and six ceiling fans. It can accommodate
a maximum of 15 people. The lab is a space which is always open and
occupants are present almost throughout the day.

4.8.4.1 *Thermal Properties of the Space*

- *Non-uniform thermal profile*: Thermal profile of the lab space, is in-
  herently non-uniform. Hot and cool regions exist in this space, due to
  the effect of glazing and wall-sharing with other work spaces. Also, the
  presence of data servers in the space adds heat to the space.
- *Thermal gradient*: IDUs tend to cool the area right under the ducts
  faster than the rest of the space. This thermal gradient is observed at
  occupant seating locations and since the seating arrangement is fixed,
  the occupants sitting under the ducts feel cool and the rest of them feel
  hot, simultaneously.
- *Influences of ambient*: The influences of the ambient are significant in
  this space, due to operable windows. The controller needs to decide
  the time instants when the windows need to be operated.

4.8.4.2 *Interventions and their Benefits*

The following interventions are applied in sequence, depending on the
present temperature and humidity values, to achieve thermal comfort.

- *Fans*: Combined use of fans and HVACs at higher set-points helps in
  uniform cooling of the space. This effect can be sustained for a longer
  period of time, as sweeping (cooler) air helps maintain thermal balance
  at the occupant seating level (refer to Figure 4.24). Fans can be used
  independent of the HVACs in this space.

Temperature profile with HVACs ON

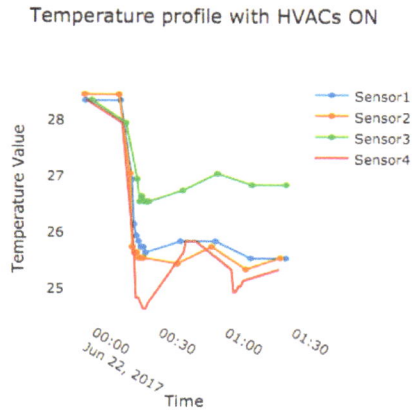

Fig. 4.22   Lab temperature profile when both IDUs are ON.

Humidity profile with HVACs ON

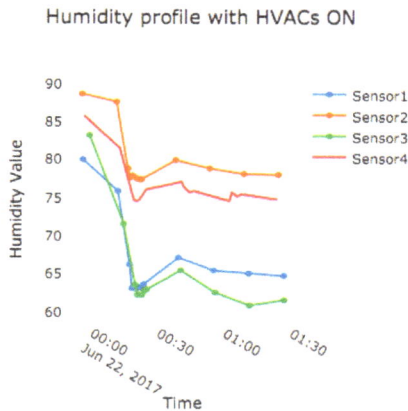

Fig. 4.23   Lab humidity profile when both IDUs are ON.

- *Operable windows*: Operable windows are an energy-efficient means to provide thermal comfort to the occupant [de Dear and Brager (2002)]. The controller needs to know when to operate the windows so that good quality air is mixed with the indoor air, periodically.

Sequence of interventions: Depending on the temperature and humidity levels at a time instant, the controller decides necessary interventions. Note that the temperatures are maintained at higher value in spaces where operable windows are used when compared to spaces without operable windows.

Temperature profile with one HVAC & Fan ON

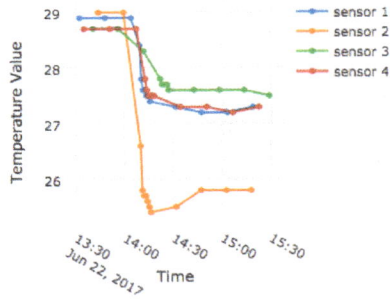

Fig. 4.24   Lab temperature profile of the space when one IDU and fans are in use.

Humidity profile with 1 HVAC & Fans ON

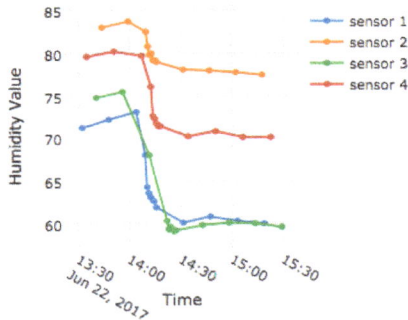

Fig. 4.25   Lab humidity profile of the space when one IDU and fans are in use.

Since humidity decides the temperature feel-like in a space, it is maintained within 10% of the lower value obtained when both the HVACs are turned ON at $24°C$ (refer to Figure 4.23).

The sequence of the interventions is as follows:

- If the humidity and temperatures are within limits, the temperature of the space can be allowed to be slightly warm. For this, just the air circulation in the space, is sufficient.
- If the temperatures are rising, while the humidity is still within acceptable limits, the number of HVACs ON can be cut down and only

one HVAC along with fan circulation is enough to bring the required thermal comfort.

• If only the humidity needs to be contained, it can be done in two ways.

    – If the outside air is dry, then we use the natural ventilation to dehumidify the space.

    – If the outside air is humid, we use HVACs and fans to dehumidify the space.

• When the humidity and the temperature levels tend to rise towards the upper limits of the acceptable ranges, both HVACs are switched ON to reduce these values.

Benefits of Interventions: To demonstrate this approach, we consider the case of lab occupancy during mid-April in a warm, humid climate zone for the evaluation of the above mentioned interventions, when the lab occupancy was about 10–12. We inspect the performance of thermal comfort algorithm between 14 hours and 18 hours. The mean of four temperature and humidity values from nodes located close to occupant seating regions, are used at run-time control.

This approach is compared with the general ad-hoc procedure of operating an appliance, at one of the seating locations in the space. In Figure 4.27, it can be observed from the occupant level Humidity and Temperature values, that warmer temperatures and therefore, higher thermal sensations (predicted on ASHRAE scale) [Brager and de Dear (2001)] are set in the lab, due to the sequence of interventions.

However, on comparing these humidity, temperature and the predicted thermal sensation scale plots, with those in the thermal profile of general usage (Figure 4.26), the following benefits can be observed:

• The spread of humidity and temperature values in Figure 4.27 are representative of the alternations between air cooling, mixing and circulation. On the other hand, in Figure 4.26, we observe clustered values, which are representative of an improper set of interventions in the space.

• The spread of the predicted thermal sensation values, in Figure 4.27 is representative of the various temperatures that an occupant is exposed to when the right sequence of interventions is applied. This indicates a prioritization of thermal preference over thermal sensation in such spaces [de Dear and Brager (2002)].

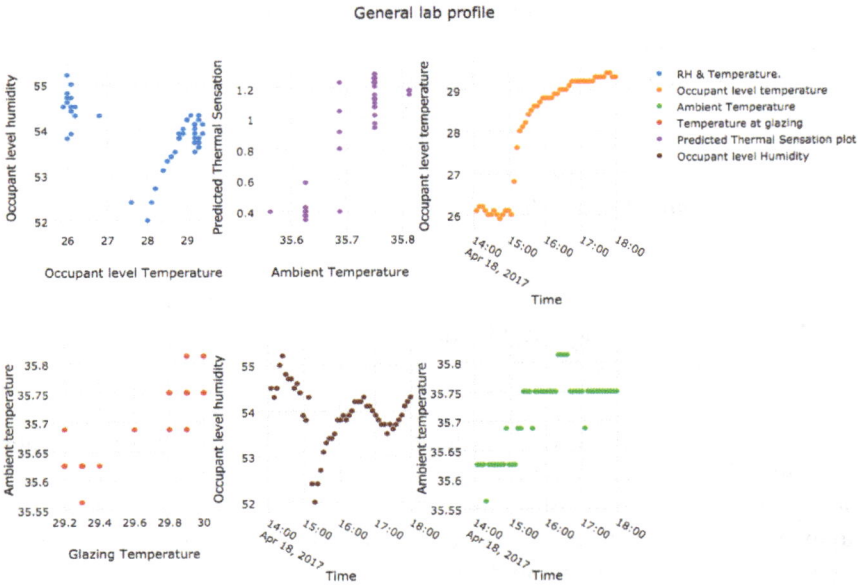

Fig. 4.26   Different observations without intervention for maintaining thermal comfort in lab.

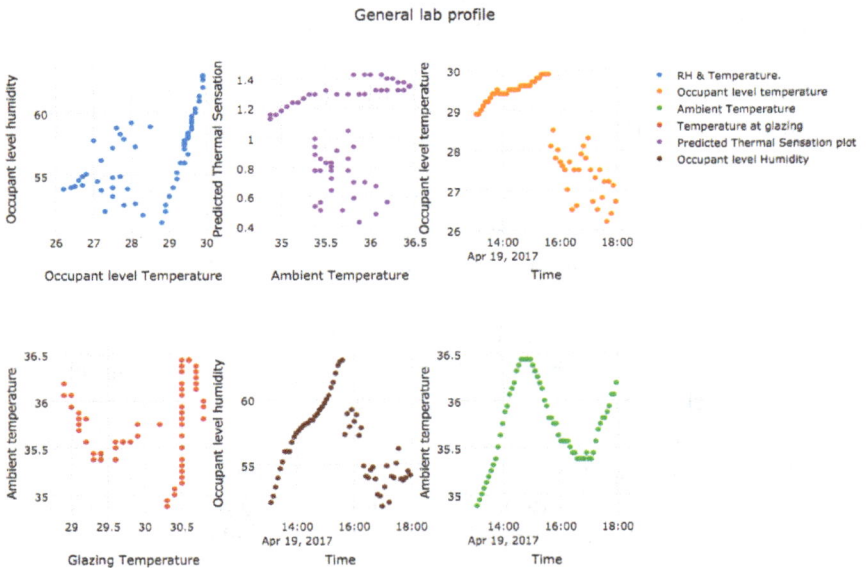

Fig. 4.27   Observations with intervention for maintaining thermal comfort in lab.

4.9   Summary and Takeaways

This chapter has presented a unified and principled approach to decide how thermal comfort in building spaces should be provisioned. We begin with examining the efficacy of the prevalent techniques for provisioning of thermal comfort and discuss thermal modeling of closed spaces in a building and an adaptive hybrid approach in thermal modeling of spaces involving ACs. We examined the decisions taken based on architectural aspects of the buildings, impact of the design decisions on providing thermal conditioning resources and run-time intervention techniques to achieve the goal.

This holistic approach is also applied in maintaining thermal comfort in real-life applications in four different types of spaces based on their sizes and provisions for conditioning resources and offered possible interventions to handle undesirable scenarios affecting the thermal comfort of the occupants. Our empirical observations demonstrate that with this principled approach, better thermal comfort can be provided with reduced consumption of energy compared to the conventional approach.

Further, adaptive demand response (D-R) techniques are introduced to meet the peak demand constraint and limiting consumption by i) scheduling TCED loads and ii) shifting/relaxation of comfort band, respectively.

Chapter 5

Customized Thermal Comfort

People are not alike and their levels of thermal comfort under the same environmental conditions are also different. However, existing systems aim to provide thermal comfort to most number of people. This is usually based on the average response of the people. It has also been observed that a centralized air-conditioning system often maintains some ad-hoc level of temperature based on experience or prevailing practice. Further, the requirement of cooling/heating of a space varies with the occupancy level and also it is not uncommon to find HVAC system running to maintain the temperature of spaces even when unoccupied. In addition, it has also been observed that in most cases, existing provisioning of HVACs fail to offer uniform cooling throughout the entire space.

Therefore, customization is required to maximize the thermal comfort level of individuals and reduce wastage of energy. In this chapter, we discuss the techniques of i) providing customized thermal comfort to individuals, ii) occupancy-based HVAC control and iii) data-driven chiller sequencing to save energy and provide thermal comfort efficiently.

5.1 Individual Thermal Comfort

Since thermal comfort is based on human cognition, it is natural that when asked, human beings express their level of thermal comfort in qualitative terms — comfortable, warm, hot, cold, very cold, etc.

In order to arrive at a quantitative response ASHRAE suggests the collection of average thermal response of a large group of subjects in terms of 7-point thermal sensation index as shown in Table 5.1, which is called *predicted mean vote* (PMV).

Table 5.1 7-point ASHRE thermal scale.

Vote	Thermal Comfort Level
+3	Hot
+2	Warm
+1	Slightly Warm
+0	Neutral
−1	Slightly Cold
−2	Cool
−3	Cold

5.1.1 Predicting Thermal Comfort

The Standard ISO 7730 [ISO (2005)] offers quantitative methods for predicting the general thermal sensation and degree of discomfort (thermal dissatisfaction) of people exposed to moderate thermal environments. It enables the analytical determination and interpretation of thermal comfort by calculating PMV and PPD (predicted percentage of dissatisfied) and local thermal comfort criteria (e.g., draught, vertical air temperature difference, etc.), giving environmental conditions that can be considered acceptable for general thermal comfort as well as for those representing local discomfort.

5.1.1.1 Predicted Mean Vote (PMV)

Studies have been conducted to compare the PMV responses directly with the measured data. The seminal work of Fanger [Fanger (1973)] since 1970 has been instrumental in the development of international standards to predict the thermal sensation of occupants [Parsons (2003)].

Evaluation of PMV: PMV can be evaluated using a function $pmv(\mathbf{x})$ that assigns a comfort value based on a vector $\mathbf{x}$ with six elements as identified by Fanger, which is already mentioned in Chapter 3.

$$\mathbf{x} = [t_a, \bar{t}_r, v_{air}, h_a, M, I_{clo}]^T$$

where

- t_a is the air temperature,
- t_r is the mean background radiant temperature,
- v_{air} is the relative air velocity,

- h_a is the humidity level,[1]
- M is the metabolic rate[2] of a person (activity level),
- I_{clo} is the clothing insulation[3] (thermal resistance) factor of a person.

When factors like environmental parameters, physical activity and clothing mentioned above are estimated or measured, the thermal sensation for the body as a whole can be predicted by calculating the PMV using the quantitative techniques in ISO 7730 Standard [ISO (2005)], which is reproduced in Appendix D.

5.1.1.2   *Predicted Percentage Dissatisfied (PPD)*

The PMV represents the mean value of the votes of the thermal sensation (in terms of 7-point ASHARE scale) of a group of people under the same environment. Since PMV is the mean value, individual votes are distributed around this value and therefore, it is necessary to predict the number of people who are thermally dissatisfied.

The index PPD is a quantitative prediction of the percentage of people who are not thermally satisfied, i.e., whose votes are other than 0 (neutral) on the ASHARE thermal scale Table 5.1.

Once the PMV value is determined, the PPD is calculated using the following equation [ISO (2005)].

$$PPD = 100 - 95 \times exp(-0.03353 \times PMV^4 - 0.2179 \times PMV^2)$$

5.1.1.3   *Acceptable Temperature and Design Criteria*

The design value of acceptable temperature of spaces can be grouped based on the activity levels. Considering i) humidity level of 50%, ii) clothing level of 0.5 *Clo* during summer and 1 *Clo* during winter, and iii) air velocity of about $1m/sec$, the acceptable operative temperature of a category of space belonging to classroom/office/conference room/auditorium/cafeteria is found to be as follows.

$$24.5 \pm 1°C \quad \text{during summer, and} \quad 22 \pm 1°C \quad \text{during winter}$$

Note that the above thermal comfort level is considered as *Category A*, which corresponds to PPD of $< 6\%$. Details on how these values are arrived

---

[1]Relative humidity (RH) is the ratio of the partial pressure of water vapor to the equilibrium vapor pressure of water at a given temperature.
[2]In met-units: 1 met = $58W/m^2 (50KCal/m^2h)$ corresponding to sedentary activity.
[3]In clo-units: $1clo = 0.155m^2 \text{deg} C/W (0.18m^2 h°C/KCal)$.

at based on PMV and PPD votes can be found in ISO 7730 standard [ISO (2005)].

5.1.2 *Predicted Personal Vote (PPV) Model*

Predicted Personal Vote (PPV) [Gao and Keshav (2013a)] model is an extension of PMV model with a provision to capture personal preferences.
PPV function $ppv(x)$ is represented as

$$ppv(x) = pmv(x) + personal(x) \qquad (5.1)$$

where $personal(x)$ models how a particular user is different from an average person and $pmv(x)$ is the output of PMV model that can be obtained using the quantitative techniques in ISO 7730 [ISO (2005)].

$personal(\mathbf{x})$ can be modeled as a linear function:

$$personal(\mathbf{x}) = \mathbf{a}^T\mathbf{x} + b$$

where $\mathbf{a}$ is the vector that models users' sensitivity to each of the six elements influencing thermal comfort

$$\mathbf{a} = [a_t, a_r, a_{air}, a_h, a_M, a_{clo}]$$

and b denotes thermal preference of the user — positive value if a user prefers colder temperature and negative value if the preference is for warmer temperature.

For training the PPV model, the measured environmental variable $\mathbf{x}$ and individual user's personal vote upv are used. upv represents the personal vote of an individual user at any point of time as calculated using Equation (5.1). From a set of $\mathbf{x}$ and upv data, the parameters $\mathbf{a}$ and b can be estimated using linear regression. In the absence of training data or sufficient data points (to do linear regression), a simple linear function $g(.)$ is trained by least square approximation to estimate PPV such that,

$$ppv(\mathbf{x}) = g(pmv(\mathbf{x}))$$

5.1.2.1 *Measuring Constituent Parameters of Thermal Comfort*

Appropriate sensors will need to be installed in order to measure the constituent parameters of thermal comfort — four environmental, one physiological and clothing insulation.

Air Temperature (t_a), relative humidity (H_a) and air flow (v_{air}) — can be measured using the combined sensors available nowadays.

Table 5.2   Metabolic Rates (ISO Standard 7730:2005).

Activity	Metabolic Rate (W/m^2)	
Reclining	46	0.8
Seated, relaxed	58	1.0
Sedentary activity (office, dwelling, school, laboratory)	70	1.2
Standing, light activity (shopping, laboratory, light industry)	93	1.6
Standing, medium activity (shop assistant, domestic work, machine work)	116	2.0
Walking on level ground:		
2 km/h	110	1.9
3 km/h	140	2.4
4 km/h	165	2.8
5 km/h	200	3.4

Background radiant temperature — can be measured using infrared thermometers which are capable of measuring a broad spectrum from very low to very high temperature.

Activity Level M — Values for typical activities are available in IEC 7730 standard, which is reproduced in Table 5.2.

Clothing Insulation I_{clo} — Values for typical clothing level are also available in IEC 7730 standard for both daily wear clothing and work clothing as reproduced in Table D.1.

But, estimation of clothing insulation requires knowledge of the set of clothes one is wearing. However, skin temperature may be assumed to be relatively constant.

The mean skin temperature t_s during comfort is formulated in [Fanger (1973)] as

$$t_s = 35.7 - 0.0276M °C$$

where M is the metabolic rate per unit body surface area (W/m^2). Therefore, during sedentary activity $M = 58W/m^2$, the preferred skin temperature is approximately $34°C$, while the preferred skin temperature is only approximately $31°C$, at an activity three times that of sedentary level.

The greater the level of clothing worn, the greater the degree of insulation, and the lower the temperature of the outermost layer of clothes. Thus, assuming constant skin temperature, the clothing insulation can be estimated [Gao and Keshav (2013a)] by measuring the temperature of clothing using non-invasive infrared thermometer. A regression model can be built to estimate the clothing insulation I_{clo} as

$$I_{clo} = f(t_c)$$

where $f(.)$ is a linear function and t_c is the temperature of the clothing measured by infrared thermometer. The model can be trained using the data set of I_{clo} estimates presented in Appendix D.1, which is reproduced from ISO standard 7730.

5.1.2.2   *A Case Study on Personalised Thermal Comfort*

Predicted Personal Vote (PPV) model (an extension of the PPV model) discussed in Section 5.1.2 was evaluated in real-world office spaces at the University of Waterloo, Canada [Gao and Keshav (2013a)]. The case study established a fairly good accuracy of estimation of thermal comfort level based on the PPV model.

The primary heating was provided by the regular HVAC system to maintain a set temperature common to all users. In addition, small radiant heaters were used in close proximity to individual users, which were controlled based on the users' personal preference of thermal comfort. Use of low cost table fans in place of heaters was suggested for warm climatic condition. The PPV estimation was used when occupancy was detected. It was observed in [Gao and Keshav (2013a)] that most of the time during occupancy, the PPV level was maintained at 0, indicating the desired thermal comfort level as denoted by the 7-point ASHRAE scale.

In order to achieve maximum personalised thermal comfort, i) feet warmers, palm warmers, etc. during cold climatic condition and ii) air-jets, hand cooling devices etc. during warm climatic condition, can also be used.

However, personalised thermal comfort discussed above can be intrusive, expensive and challenging to implement in a large scale setup.

5.2   Occupancy Based Customization

The simplest rule in saving energy by detecting occupancy is — *Switch OFF When Unoccupied* i.e., switch OFF the corresponding HVAC appliance or actuator whenever any thermally isolated zone is not occupied [Agarwal *et al.* (2011)]. Implementing this rule requires i) sensing of occupancy in the identified thermal zones where HVACs can be controlled and ii) identification and automatic control of HVAC equipment, (in case of room air-conditioners) or actuators like Variable Air Volume (VAV) air handler (in case of centralized system).

However, in order to deal with varying occupancy levels, the control scheme can be complex, demanding and it can involve actuator controls beyond the usual ON-OFF control.

5.2.1 *Occupancy Sensing*

Detection of occupancy is a significant component in any BMS and plays a very important role in energy management. However, this is a challenging task considering the fact that the various available techniques have their own merits and demerits. There is no "one size fits all" solution in occupancy detection. Further, it is not practical to deploy a very large number of physical sensors for this purpose, as this can affect the aesthetics of the existing buildings and introduces challenges in maintenance. Therefore, recent works propose a fusion of sensors and judicious exploitation of existing infrastructure like smart doors and wi-fi access points.

5.2.1.1 *Fusion of Sensors for Occupancy Sensing*

It has been discussed in Chapter 3 that no single occupancy detection technique is capable of handling the sensing needs in buildings. This is because, cost, accuracy, network bandwidth requirement and also user privacy play a big role in real-world implementations. Fusion of more than one sensor is the key to overcome the disadvantages of one-size-fits-all solution as discussed in Section 3.3.3 of Chapter 3.

5.2.1.2 *Exploiting Wi-Fi System for Occupancy Sensing*

The occupancy detection techniques discussed so far require deployment of physical sensors. Deployment of physical sensors involves additional cost towards installation and maintenance. Furthermore, it is likely to impact the aesthetics of existing buildings.

This has led to the exploration of possibilities that involve existing Wi-Fi infrastructure in occupancy sensing. Recent works show that existing Wi-Fi network in combination with Wi-Fi enabled smartphones of the occupants can be utilized to infer occupancy in a building. However, such attempts are often a trade-off between cost of deployment (of physical sensors) and detection accuracy. It has been shown in [Balaji *et al.* (2013a); Ardakanian *et al.* (2016a)] that it is feasible to determine occupancy in an existing building from the readily available information from the installed Wi-Fi access points.

5.2.1.3   Occupancy-Based HVAC Control

The options for occupancy-based HVAC control depend on the type of system deployed in a building irrespective of the occupancy-sensing methods used. Based on the types of HVACs, energy can be saved by controlling the i) compressor of the AC unit, ii) in-door unit(s) (IDU) and iii) variable air volume (VAV) unit(s) deployed in the spaces where occupancy can be sensed.

Compressor control: This is applicable where stand alone HVAC unit(s) are deployed in thermally isolated spaces for controlling the thermal environment.

IDU Control: In case of VRF (Variable Refrigerant Flow) HVACs, the IDUs in the individual spaces can be switched ON or OFF depending on the occupancy state. The compressor of a VRF HVAC adjusts the flow of its refrigerant depending on the thermal loads derived from the number of IDUs that are ON and their temperature set points.

In this context, it is to be noted that in many cases, where a relatively large number of spaces are involved, VRF systems deploy more than one compressor. In such cases, at least one compressor is not provided with refrigerant flow control device assuming a base load demand, i.e., a minimum number of spaces are always assumed to be occupied, requiring maintenance of thermal comfort. Therefore, a limit is introduced where no further energy can be saved by switching OFF more IDUs beyond a certain number.

VAV control: Use of Variable Air Volume (VAV) systems is very popular in many modern buildings, as it offers improved energy efficiency and better control on thermal comfort level. VAV systems are suitable for a centralized HVAC as they often involve many zones with diverse airflow needs and dynamic occupancy levels. VAV control units are deployed in the air handling (AHU) section of a centralized air-conditioning system. Each zone has a VAV unit, often referred to as VAV box or terminal. This unit can i) modulate air flow based on cooling load and ii) maintain minimum airflow for ventilation requirements. It also has the capability of reheating air, when needed.

Therefore, in a VAV system there is scope for saving energy by controlling airflow based on the occupancy level and not just ON-OFF control.

Chiller Sequencing: Use of multiple chillers is common in centralized cooling systems of large commercial buildings. The number of chillers which are to be put into operation at any point of time depends on the cooling load, which is affected by occupancy level at different times of the day. The varying cooling load calls for chiller sequencing in real-time, which is performed by determining and operating the most energy-efficient combination of chillers.

5.2.1.4 *Schedule Driven HVAC Control*

Each building's heating and cooling is controlled by a commercial HVAC system. Unlike a residential HVAC system, which is controlled by a thermo-stat, a commercial HVAC system is typically controlled through a Building Management System (BMS). The building's facility manager interacts with the BMS to set a heating and cooling schedule and temperature setpoints: this schedule specifies when the HVAC equipment should be turned on over the course of a day and the temperature setpoints for the high and low occupancy periods. The BMS then automatically operates the building's HVAC equipment based on the specified schedule, a process we refer to as *schedule-based HVAC control.*

The schedule is typically determined based on the facility manager's intuitive understanding of the building's occupancy patterns. For instance, in a typical office environment, employees may arrive between 8 am and 9 am and leave for home between 4 pm and 5 pm, and the building may be lightly occupied during non-business hours or on weekends. In this example, the facility manager may program the BMS to heat or cool the building between 8 am to 6 pm and use a higher cooling or lower heating temperature during the off-peak hours. By doing so, it is ensured that users are comfortable when the building is highly occupied, while saving energy when it is largely unoccupied.

Modern BMS systems enable a different schedule to be set in different parts of a building — e.g., a different schedule on different floors — if the per-floor occupancy patterns differ. However, to fully exploit this function-ality, a facility manager needs to determine fine-grain schedules for differ-ent parts of a building, and dynamically fine-tune it as occupancy patterns change over time. Such manual operation is cumbersome and error-prone and does not scale across a large campus, where a facility manager may oversee tens-to-hundreds of buildings. As a result, existing schedule-based HVAC control tends to be driven by simple, manually-chosen static sched-ules, which miss many opportunities for reducing energy use by carefully

(a) Schedule-based HVAC control

(b) Occupancy-based HVAC control

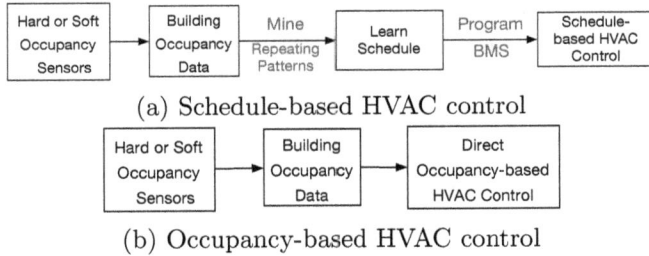

Fig. 5.1   Schedule versus occupancy-based HVAC control. (Courtesy [Trivedi *et al.* 2017])

exploiting temporal and spatial occupancy differences within and across campus buildings.

To address these limitations, the process of deriving schedules for commercial HVAC systems should be *automated*. To do so, the system needs to monitor occupancy patterns in campus buildings, automatically learn an optimal schedule for each part of a building based on the observed occupancy, and dynamically modify the schedule as occupancy patterns change. Such a system should be sufficiently robust to tailor its schedules to the different types of *spatial* occupancy patterns seen in different types of campus buildings, e.g. classrooms, academic units, library, dining halls, on a university campus. It should also automatically tailor schedules for *temporal* variations seen across weekdays and weekends and across seasons.

Figure 5.1(a) depicts the architecture of such a schedule-based HVAC control system — hard or soft sensors are first used to infer occupancy in each building. Occupancy data is then "mined" to extract repeating patterns, from which an "optimal" schedule is learned and fed to the BMS for schedule-based control of the commercial HVAC system.

An essential first step in such a learning-based system is to derive occupancy, which captures *how many* people are present in each part of a building and at *what times*. Hard sensors such as motion or door sensors can be used to track occupancy within each building [Beltran *et al.* (2013); Agarwal *et al.* (2010); Newsham and Birt (2010); Tapia *et al.* (2004)]. However, such instrumentation is not ubiquitous in office buildings and can be expensive and laborious to install in existing buildings. Researchers have shown that occupancy can also be learned through "soft sensors" that are already deployed for other reasons. For example, occupancy can be learned through swipe card door access systems, calendar software, or through wireless network activity [Ghai *et al.* (2012b); Melfi *et al.* (2011); Ting *et al.* (2013);

Milenkovic and Amft (2013)]. Since WiFi infrastructure is now ubiquitous in offices and campus buildings, existing wireless networks rather than hard sensors can be used to infer occupancy information. Doing so enables easy deployment of such a system in today's campuses without requiring expensive deployment of hard sensors.

Specifically, most occupants carry mobile smartphones and the presence of a phone in the vicinity of a wireless access point indicates a user (occupant) at that location. The exact location of each access point within a building is assumed to be known a priori. Consequently, simply tracking the number of mobile devices associated with each AP over time is a proxy ("soft sensor") for the number of occupants in that part of the building. The wireless network infrastructure is assumed to provide a log of when a mobile device connects and disconnects to each access point, which is then used to count the number of active occupants over time. To avoid double counting users, only smartphone log entries are considered and other devices, such as laptops or stationary devices are filtered out.[4]

Problem Statement: Given an office campus with multiple buildings, anonymized WiFi association and dissociation logs, and the mappings of the Access Points to specific locations in each building, our goal is to automatically learn an HVAC schedule that optimizes user comfort and energy usage on a fine-grain spatial basis, and dynamically adjusts learned schedules when observed occupancy patterns change.

Prior Work

Occupancy monitoring and optimizing HVAC efficiency has received significant attention in recent years. This section describes the differences and specific contributions of our work in relation to these past efforts.

Occupancy-driven versus schedule-driven HVAC control: Efforts such as Sentinel and others [Balaji *et al.* (2013b); Agarwal *et al.* (2010)] have shown how occupancy sensors can *directly control* HVAC systems. The basic approach, depicted in Figure 5.1(b), uses observed periods of high and low occupancy to directly control HVAC systems and save energy during off-peak periods. This approach, while novel, is not compatible with most existing BMSs that employ *schedule-driven* control (Figure 5.1(a)). In the latter approach, occupancy data is first used to learn a repeating schedule,

---

[4]Note that the system only needs to count the number of active devices at an AP and does not need to track individual users — identifiable information such as device MAC addresses can thus be anonymized.

which is then set in the BMS to control the HVAC system. Thus, occupancy information only indirectly, rather than directly, influences HVAC operation.

While direct occupancy-driven control approaches may be appropriate for buildings with local (e.g., room-specific) HVAC units [Karmakar *et al.* (2015b)], they are not viable for the majority of centralized commercial HVACs controlled through BMS schedules. In addition, given their experience with schedule-based control, many facility managers may be uncomfortable with ceding direct HVAC control to software. Thus, by deriving repeating occupancy-based schedules, we enable facility managers to retain some measure of control over HVAC usage.

Residential versus Commercial: In residential settings, efforts such as smart thermostat [Lu *et al.* (2010)]; iProgram [Iyengar *et al.* (2015)], as well as products such as Nest, Ecobee and Lyric, have been used to improve HVAC energy-efficiency. Such smart thermostats, as well as all "dumb" programmable thermostats, use schedule-based HVAC control, where occupancy information (from onboard sensors, phone GPS, or even electricity meters [Iyengar *et al.* (2015)]) is analyzed to automatically learn a custom schedule. Occupancy sensors may occasionally turn on "away" mode, but they do not exercise direct control. User feedback has also been used to optimize HVAC use [Gao and Keshav (2013b); Lam *et al.* (2014)]. While homes need only binary temporal occupancy, larger commercial buildings need spatial occupancy data. Thus, our work can be seen as analogous to these residential efforts but applied to commercial buildings — a more complex problem.

Inferring Occupancy: There has been significant work in deriving occupancy information both for residential and office buildings. Prior work on deriving occupancy information falls into three categories: (i) design of novel occupancy sensors, (ii) use of existing soft sensors [Prakash *et al.* (2015)], and (iii) use of energy analytic methods to learn occupancy [Pisharoty *et al.* (2015); Beltran *et al.* (2013); Lam *et al.* (2014); Gao and Keshav (2013b); Clear *et al.* (2014); Agarwal *et al.* (2010); Milenkovic and Amft (2013); Erickson *et al.* (2011); Kleiminger *et al.* (2014, 2013b)]. However, most approaches only derive occupancy and do not apply it for HVAC control. As shown in Figure 5.1(a), deriving occupancy data is only a necessary first step for smart HVAC control and is not sufficient for addressing the broader control problem. One closely related technique combines soft sensing with HVAC scheduling [Ardakanian *et al.* (2016b)]; human occupancy is sensed by monitoring human-induced HVAC heat loading and

Table 5.3 Comparison of various approaches. [Trivedi *et al.* (2017)]

System	Derive Occupancy	Learn Schedules	HVAC Control
Occupancy sensing efforts	Hard or Soft Sensors	None	None
Sentinel [Balaji *et al.* (2013b)]	WiFi-based soft sensors	×	occupancy-based
Ardakanian *et al.* [Ardakanian *et al.* (2016b)]	HVAC-based sensors	ML-based	schedule-based
iSchedule	WiFi-based soft sensors	ML-based	schedule-based

is used as feedback to modify an existing schedule. While our system, iSchedule, also derives HVAC schedules, it instead leverages WiFi-based soft sensors for predicting occupancy. WiFi-based soft sensors can more explicitly derive occupancy counts since there is a direct mapping between numbers of device associations and numbers of occupants. In contrast, our system is easily deployed across an entire campus rather than relying on more specific feedback from advanced HVAC functionality, which could be limited to more recently constructed buildings with newer HVAC units. Furthermore, HVAC-based soft sensors can only operate at the granularity of already defined zones; WiFi access points are typically deployed at a higher spatial density throughout a building, enabling a building manager with data needed to potentially redefine zones in the future.

In order to exploit the existing schedule driven control, a new approach called iSchedule is proposed in [Trivedi *et al.* (2017)] where, occupancy data is first used to learn a repeating schedule, which is then set in the BMS to control the HVAC system. Thus, occupancy information influences the HVAC operation indirectly, rather than directly. In this approach, a machine learning driven technique is used to automatically learn custom occupancy-based HVAC schedules for buildings across a large campus.

It is assumed that the system receives a raw log of smartphone association and disassociation information to each access point in the wireless network. Practically, every commercial enterprise's wireless network products routinely log such information (e.g. Cisco, HP Aruba). The location of each access point within each building is assumed to be known.[5] Given this data, iSchedule learns schedules as follows:

---

[5]Since a user may own multiple mobile devices, double counting is avoided by only counting mobile phones connected to an Access Point and ignoring other device types.

Step 1: Compute Temporal Occupancy Per Access Point
The system processes the raw WiFi logs to partition the logs on a per access point basis. It then computes the number of active devices (i.e., users) connected to the Access Point (AP) in each time interval. This is done by incrementing the number of active users upon each new device association and decreasing it for each disassociation event. Doing so yields the number of active users in the vicinity of that AP during each time interval (e.g., every 15 minutes or hourly) over the duration of the log.

Step 2: Derive Spatial Occupancy within a Building
Since the location of each AP in a building is known, we can group all AP's spatially to obtain the observed occupancy within each part of the building. Any spatial grouping can be chosen (depending on how fine-grained the HVAC control can be). The default grouping is on a per-floor basis. By aggregating the temporal occupancy seen by all APs on each floor, the number of users that are present on that floor in each time interval over the duration of the WiFi log are obtained. This yields a spatial distribution of users across the building that captures the change in spatial occupancy over time.

Step 3: Use Predictive Model to Infer Floor/Zone Occupancy
Next, the system predicts the occupancy of each floor/zone. A supervised training technique is used to predict occupancy. A Gradient Boosting Regressor Ensemble model is trained using the occupancy data computed in the previous step. In the case of university campus buildings, the following features are a strong indicator of occupancy and form our feature set: (i) building name, (ii) building floor or spatial region, (iii) day of the week, (iv) time interval (e.g. hour of day or a 15 minute interval, 9:00 AM to 9:15 AM, etc.), (v) semester of the year (vi) month, (vii) holiday and (viii) year. The floor/zone occupancy forms the label set.

Dynamic Adaptation of Learned Schedule
Occupancy patterns are not stationary and will slowly (or abruptly) change over time. These changes in occupancy patterns may occur for a number of reasons: the building or floor may get re-purposed for a different class of users. For example, an academic building may become administrative space with new types of users moving in or there may be subtle changes in occupancy patterns with different types of users over time (e.g., due to changing of class schedules or different user patterns). Regardless of the cause, the learned schedules cannot remain static — they must adapt and evolve with changing occupancy patterns. In other words, once learned, the HVAC schedule must be dynamically and periodically recomputed and

adjusted. The algorithm presented in the previous section can be enhanced in one of two ways to support adaptation.

Continuous Adaptation: In this method, WiFi activity data is ingested every day and spatial occupancy observed within each building during that day is added to the historical trace. The predictive model is relearned using all data, including the newly ingested information, and the HVAC schedule (step 5) is re-computed. The frequency with which the schedule is recomputed is made configurable (e.g., daily, weekly, monthly, etc.).

On-demand Adaptation: A limitation of the continuous adaptation approach is that it wastes computational resources when no significant changes to occupancy are observed, as the model is retrained periodically, regardless of whether it is necessary. On demand adaptation is an alternate approach that triggers retraining only when the prediction deviates from observed occupancy. As before, new WiFi activity data arrives continuously and is added to the historical data repository. The system then periodically invokes the previously learned predictive model to predict high and low occupancy labels for a recent time interval. The model predictions are compared to the actual occupancy levels observed in the newly captured WiFi-based occupancy data. If the model predictions match the observed levels, then the occupancy patterns are the same as before and neither the model nor the HVAC schedules need to be adjusted. On the other hand, if the recently observed occupancy levels begin deviating from model predictions, then our system triggers a retraining of the predictive model and uses the new model to recompute the HVAC schedules. Thus, a new model is learned only when needed and only for those buildings (or parts of a building) where significantly different occupancy patterns are observed. The threshold error ϵ between model predicted and actual observations that trigger a relearning is configurable: a smaller ϵ triggers more frequent recomputations and schedule adjustments and vice versa.

5.3 Chiller Sequencing — Customization for Varying Loads

The goal of a chilling plant is to provide thermal comfort most economically. Usually, more than one chiller are deployed in order to provide thermal comfort to a large number of spaces/buildings with varying cooling loads. This is because, the performance of a chiller varies with changes in cooling load. Note that cooling loads vary i) due to varying occupancy or ii) variation in ambient temperature due to changes in season. In order to ensure economic operation, multiple chillers are operated in such a way that they achieve the highest possible overall coefficient of performance (COP). Before discussing further, let us understand the term COP.

Coefficient of Performance (COP), is a measure of the energy-efficiency of chillers (or space heaters and other cooling and heating devices). COP (BTU/W) is equal to the heat extracted/delivered (output) in British thermal units (BTU) per hour divided by the heat equivalent[6] of the electric energy input. Thus, in case of a chiller, it is the ratio of heat extracted in the evaporator (Figure 4.2) to the work done (in watt) by the electricity-driven compressor. Higher COP indicates higher efficiency of a chiller and therefore, lower operating cost.

This calls for chiller sequencing control, which determines how many and which chillers are to be put into operation in accordance with the instantaneous building cooling load demand. The overall COP is derived taking into account the COPs of individual chillers at different loads, typically 25, 50, 75 and 100%, under different operating conditions. A chiller can operate with its cooling load varying from 30% to 100%, but considering its COP, it is preferred that no chiller operates at a load below 70%. The goal of the chiller sequencing control is to maximize the overall COP of the chillers over a typical cooling period or season.

5.3.1   *Chiller Control Techniques*

Industrial practice that is followed in the chilled water circuit is to supply water at a temperature T_{out} of about 7.5°C and the return line water temperature T_{in} is maintained at about 12.5°C. The differential temperature ΔT is maintained at about 4°C. The cooling load Q_i is determined by

$$Q_i = m_i C \Delta T \qquad (5.2)$$

where, C (kJ/kg °C) is the thermal capacity of water and m_i kg/s is the chilled water mass flow rate.

If cooling load increases, water in the chilled water circuits extracts more heat from the spaces and therefore, the return water temperature T_{in} increases. In order to match the increase in load, more heat is extracted from the refrigerant (at evaporator) in the chiller circuit and the load on the compressor increases. If the cooling load decreases, it causes decrease in T_{in} and a reduction in the compressor load. In other words, the demand of electrical energy by the compressor varies (increases/decreases) according to the changes (increase/decrease) in the cooling load. Thus, both return

---

[6]1 watt (W) = 3.413 BTU/hr.

line water temperature T_{in} and the current I_c drawn by the compressor are indicators of changes in the load.

A chiller plant traditionally operates in accordance with a fixed schedule of load, which is determined based on the historical data of occupancy and operating equipment. One or more chillers are put into operation based on load schedule and whenever there is a change in load, the flow of the refrigerant in the chiller circuit is controlled to match the load demand. When all the running chillers reach their full load capacities, an additional chiller is brought in if the cooling load increases further. This may cause the previously operating chillers to run at a lower load due to excess capacity that may be added by the additional chiller. The accepted practice is to determine the number of chillers so that no chiller runs below 70% of its full load capacity. Following the same principle, one of the running chillers is staged off when the cooling load goes below 70% and the total cooling load is shared by the operating chillers.

Accurate determination of chiller loads along with the part load efficiencies of the chiller system components is necessary to achieve the goal of maximizing overall plant efficiency. The cooling load is determined from Equation (5.2) by measuring the ΔT and the mass flow rate. In addition to determining the number of chillers to run in accordance with the cooling load demand, it is also important to determine the share of cooling loads on individual chillers towards reaching higher overall efficiency.

The rated capacity of a chiller is the design specification, which is fixed, but the performance of a chiller is affected by a number of factors, like chiller load, ambient temperature, operating condition, performance degradation over time, etc. Therefore, various methods of sequencing chillers [Shan *et al.* (2016)] are used in practical applications viz. return chilled water temperature based control, direct power based control, bypass flow based control, etc. However, control based on cooling loads (Equation (5.2)) is most dominant as it is based on direct measurement of loads [Sun *et al.* (2013)]. Most prevalent control strategies are as follows.

- Strategy $S1$: *Conventional determination of number of operating chillers using direct measurement of cooling loads and rated cooling capacity of chiller*
  Direct measurement of cooling load Q_i and the chiller rated capacity are used in strategy $S1$ to determine the number of chillers N_c required to meet the cooling load demand at any point of time. Based on this number, staging a chiller *on* or *off* is decided.

- Strategy *S*2: *Based on chiller operating current and return line chilled water temperature*

  In this strategy, staging *on* of an additional chiller is done if the current drawn by a running chiller goes higher than a set value and it sustains for a pre-defined time. Higher chiller current I_c is an indicator of higher cooling load, but lower value of I_c is not an accurate indicator of lower demand or excess cooling. Therefore, chillers are staged *off* based on the direct measurement of cooling load and according to its predetermined set values.

- Strategy *S*3: *Fusion of cooling load measurement and online computed chiller maximum cooling capacity from a simplified model*

  This strategy employs the conventional method of determining the number of operating chillers required, as adopted in strategy *S*1. But, the techniques employed in determining the cooling loads and the rated cooling capacity are different.

  Fusion of direct and indirect measurements (using power consumption, evaporating temperature and condensing temperature) is used in determining the cooling load for improved accuracy.

  In this strategy, the rated cooling capacity, which often deviates from the actual one in an operating plant is calculated online using a model of the chiller. The online calculation using parameters representing actual operating conditions offer a better estimation of cooling capacity.

5.3.2   *Data-Driven Techniques in Chiller Sequencing*

This is a recent advancement towards improving the accuracy. In order to arrive at the performance of a chiller at various cooling loads, mainly two techniques are employed viz., *initial profiling* and *thermodynamic modeling.*
Initial Profiling: The COP profiling is undertaken during installation and commissioning stage of the chiller, for its designed operating range of cooling load.
Thermodynamic Modeling: Thermodynamic models are developed for computing COP online. This helps in capturing the variation in COP that may occur due to changes in operating conditions like evaporating temperature T_e and condensing temperature T_c of the refrigerant.

Initial profiling has the drawback of using the COP data determined during the commissioning stage, which often changes with the operating conditions of the chiller. On the other hand, the fixed form thermodynamic models have their own limitations as COP depends on various other factors, like configuration dynamics (e.g., variable flow by the chilled water pumps), varying cooling demand, degradation over time, etc. This makes it difficult for an analytical model to capture the impact of these variables accurately and makes a case for data-driven COP profiling.

Data-driven chiller sequencing involves two steps — i) Prediction of COP for individual chillers based on historical data and ii) Determination of optimal sequencing of chillers [Zheng *et al.* (2018)]. It is to be noted that in addition, successful execution of chiller sequencing requires feedback control.

5.3.2.1 *Time-Constrained Data-Driven Chiller Sequencing (TCDD-CS)*

A time-constrained data driven chiller sequencing technique is proposed in [Zheng *et al.* (2018)], which involves the following two steps.

- Data-driven COP prediction, and
- Determination of optimum sequencing of chillers.

Let us look into the practical considerations to be taken into account, before discussing further on the TCDD-CS problem and its solution.

5.3.2.2 *Practical Considerations*

Before executing any chiller sequencing, the following real-world aspects must be taken into consideration.

- Chillers cannot be staged on/off frequently like tasks, which can be executed/preempted by an operating system. Once a chiller is started, it cannot be switched off immediately and also, a minimum restart-delay must be ensured after, as the chiller is switched off. Thus, a minimum up-time and restart-delay constraints are involved in order to avoid likely damage to the mechanical components of a chiller.
- Chillers are supplied along with their own closed loop control system and any change in sequence introduces disturbance in the system. Therefore, it is necessary that sufficient time is allowed to the system to regain its stable operating state before adopting the next sequence.
- Based on the historical data of varying cooling loads and/or previous schedules, the periodicity of triggering chiller sequence is to be decided so that it does not cause unnecessary switching (staging on/off) of

chillers and at the same time provides thermal comfort with improved COP.

Therefore, coordination among the COP computation time, the chiller up-time and restart-delay is required to ensure successful execution of any chiller sequencing in a real-world plant.

5.3.2.3   *Data-Driven COP Prediction*

For a cooling demand Q^D, the first step towards obtaining an optimal (maximizing the energy efficiency) chiller sequence is inferring the COP of individual chillers from the historical data of chiller plant using machine learning techniques.

Let COP_t denote the predicted COP at any time t over a total duration of T. This is based on historical data under various combinations of operating conditions, weather, cooling demands, aging of equipment, etc. The COP profile can be inferred by minimizing the prediction loss formulated as

$$\frac{1}{T} \sum_{t<T} [(f_t(X_t, W) - COP_t)^2] \qquad (5.3)$$

where, X_t denotes the features at time t of total time T, W denotes the corresponding parameters and $f(.)$ denotes the learning and prediction models.

The features for chillers can be i) temporal — season of the year, aging of chiller, ii) meteorological — ambient temperature and humidity and iii) mechanical — chiller power (KW) rating, make or model, cooling load, mass (water) flow rate.

5.3.2.4   *Determination of Optimum Sequencing of Chillers*

The cooling load changes dynamically. Therefore, periodic checking of cooling load is necessary before determining any particular sequence of chillers. The periodicity of sequencing will depend on the following.

- The minimum restart-delay requirement of the chillers,
- The minimum time required by the feedback control system in place, so that the system regains its stability before it can accept a new sequence, and
- The COP computation time.

Determining the Number of Chillers and Staging on–Staging off Control

The thresholds Q_{on}^T for staging on and Q_{off}^T for staging off the chillers can be determined by

$$Q_{on}^T = N_0(Q_r^C + d) \tag{5.4}$$

$$Q_{off}^T = N_0(Q_r^C - d) \tag{5.5}$$

where, N_0 is the number of running chillers, Q_r^C is the rated cooling capacity and d is the dead band. Switching on threshold is kept higher than the switching off threshold for obvious practical reasons. The dead band d is user defined and this is required for two reasons; i) it is impractical to switch on and off a chiller at the same set point and ii) it must be ensured that chiller staging on and off do not occur due a small change in the cooling load.

The number of required chillers is computed as

$$N_c = \begin{cases} N_0 + 1, & \text{if } Q^D \geq Q_{on}^T \\ N_0 - 1, & \text{if } Q^D \leq Q_{off}^T \end{cases} \tag{5.6}$$

where, Q^D is the total cooling demand.

Equation (5.6) gives a simple technique to determine the number of required chillers N_c for a given load Q^D. However, before staging off any chiller, it should be ensured that the remaining chillers are able to share the additional cooling load that would be thrown off by the staged-off chiller. This will have an impact on determining the value of threshold d especially in determining the staging off threshold Q_{off}^T.

Given the required number of chillers N_c, the data-driven predicted COP of the available chillers, what we need is to determine a chiller sequence $Q = \{Q_i\}$, which will minimize the energy consumption

$$E = \sum_{i=1}^{N_c} E_i$$

where, E_i is the energy consumption of the ith chiller with a cooling load of Q_i.

5.4 Adaptive Thermal Comfort

In the previous sections, we discussed customization of thermal comfort based on i) individual thermal preferences and ii) varying cooling loads due to varying occupancy and/or varying ambient temperature. But, it is to be noted that the underlying assumption of the discussions was that human thermal preference is not influenced by previous exposure to hotter or cooler climates [Fanger (1973)], but only by the six elements discussed in Section 5.1.1.1. Also, the influence of psychological adaptation like expectation level due to regular exposure to harsh thermal environment, is not considered in [Fanger (1973)].

However, experience and behavioral research tell us that an individual's experience of a place is not just a physical phenomenon. It is a multivariate phenomena and a person's individual objectives and expectations from the place are reflected in the degree of his or her comfort level, including thermal comfort.

[Brager and de Dear (1998)] argue that

"while the factors that have been tested have been demonstrated time and again to be irrelevant to the comfort response of subjects in the contrived setting of the climate chamber [Fanger (1970)], there remains a lingering suspicion in the minds of many researchers and practitioners alike that non-thermal factors cannot be dismissed so easily"

An adaptive approach to modeling thermal comfort is based on the hypothesis which states that "one's satisfaction with an indoor climate is achieved by matching the actual thermal environmental conditions prevailing at that point in time and space, with one's thermal expectations of what the indoor climate should be like." Thus, the thermal expectations are the result of one's past experiences and cultural practices.

The adaptive model recognizes that one's perception of control makes adverse stimuli less annoying, if it is perceived that (s)he has control over them. Thus, provisions of operable windows and fans along with HVACs may offer higher degree of thermal comfort compared to a situation where the user is a passive recipient of a given thermal conditioning.

Human thermal adaptation can be through behavioral adjustments, physiological and psychological. The static heat balance model does not account for all of the following human responses to thermal stress, namely,

Behavioral adjustment of clothing and activity, opening windows, running fans and/or HVACs, humidifiers and de-humidifiers,

physiological acclimatization like increased sweating capacity for a given heat load due to exposure to heat, and *psychological* adaptation like repeated or chronic exposure to any environmental stress like heat or cold having diminutive effect on the intensity of user's response.

The adaptive model aims to capture the human being's interaction and adaptation with the environment in order to offer i) improved understanding of the influence of human adaptation on thermal comfort, ii) improved predictive models and control algorithms and iii) design of energy conscious and climatically responsive buildings.

However, work in the domain of thermal comfort is dominated by classical heat balance model with PMV [Fanger (1970)], and thermo-physiological model [Huizenga *et al.* (2001); Gagge *et al.* (1986)]. But, in order to quantify how exactly the same environment affects different individuals, it would require inclusion of all factors — thermodynamical, physiological, psychological, cultural, and contextual, which influence the thermal perception. Research in these areas is not fully developed and hence not considered further in this monograph. A comprehensive review on the drivers in diversity in human thermal perception and holistic comfort model can be found in [Schweiker *et al.* (2018)].

5.5 Summary and Takeaways

This chapter discussed the techniques for providing thermal comfort that can be customized to i) the need of the individuals and ii) meet the right demand economically under varying occupancy levels.

Personalized thermal comfort using Predicted Personal Vote (PPV) model, a recent extension of the Predicted Mean Vote (PMV) model of ASHARE, was discussed along with a case study. Occupancy-based HVAC control is presented along with various techniques of occupancy detection including a predictive model, which is dynamically adaptable to infer floor or zone level occupancy. In addition, data-driven chiller sequencing is also presented, which is a recent advancement offering improved accuracy in determining the coefficient of performance (COP) and optimum sequencing of chillers.

Chapter 6

Solar Energy in Buildings

Renewable energy sources and distributed generation (DG) technologies can offer sustainable solutions to urban energy demands. Renewable energy sources are continuously renewed or replenished by nature. The main source of renewable energy is the Sun. Renewable energy is being seen as a transformative solution to address energy as well as climate challenges. For example, there is an increasing focus on the development of solar energy in India for a variety of reasons, including our limited conventional energy reserves, their local environment and social impacts, energy security issues, energy access and tackling the challenge of climate change. Solar photovoltaics (PV) technology is emerging as an extremely attractive option, particularly with abundantly available solar resource, modular technology and zero fuel costs over 25 years of the project life. A large share of it is expected to be achieved through decentralised and rooftop-scale solar projects. Rooftop solar PV will play a prominent role in meeting energy demands across different consumption segments. It has already achieved grid parity for commercial and industrial users and is fast becoming attractive for residential consumers as well.

The most attractive property of solar energy is that it is renewable and clean. However, due to the intermittent and unpredictable nature of renewables like solar and wind, storage is required to exploit them as stable and dependable resources. Therefore, efficient energy storage mechanisms and their judicious exploitation (with respect to economic efficiency and environmental sustainability) are critical in integrating renewable resources into the grid.

Around 75% of the electricity generated in India is consumed by buildings, out of which residential consumers account for 21.79%, and commercial entities use 8.33% of the total [CEA-India (2013)]. The challenge to reduce electricity consumption in buildings is ongoing research, and varied approaches have been proposed to achieve an aggressive reduction. However, DG from renewable sources like solar PV has faced *technical and non-technical challenges* and hence its widespread adoption is prevented.

> The solar panels generate energy during the day while a typical residence has its peak consumption during the morning and night.
> Solar PV modules are often affected by partial shading because of the non-availability of direct insolation throughout the day. This is more dominant in urban areas due to shading caused by neighbouring buildings, structures (e.g., billboards, poles, etc.) and trees.

Such DG installations rely on net metering, where consumers give the excess energy generated to the grid and receive economic incentives from utilities. This makes DG less attractive in states where net metering policy is not applicable. Even in locations with net metering, utilities restrict the amount of energy that can be supplied back to the grid. This cap is helpful for utilities in maintaining grid stability and balance the supply and demand when consumers feed intermittent power from renewables into the grid.

Therefore, to effectively reduce costs, a consumer needs more than financial incentives and net metering policy for balancing on-site generation and consumption. Almost all cities in India have scheduled power-cuts due to deficit or outages due to faults in aging grid infrastructure. Thus, residential consumers require better incentives and technological solutions before they invest in rooftop solar PV systems.

6.1   Exploiting Solar Energy: Potential and Approaches

The prospects of utilizing rooftops in urban areas is not quite sunny. Due to land scarcity, much of the urban land is covered by built-up areas. The available rooftops are occupied by overhead tanks, AC exhausts, etc., leading to structural stability issues in installing solar panels. Further, due to high levels of dust particles and pollutants (vehicular and industrial) in the atmosphere, the spectral distribution of sunlight is more diffused in urban areas.

Building simulation tools are essential in analyzing and planning for solar potential in urban areas. However, conventional building simulation tools use solar geometry in Cartesian and hence require high computation time. They do not consider the future growth possibility of the neighborhood. The impact of building-integrated photovoltaics (BIPV) on human comfort has been studied only for one or two technologies, and for single orientation.

6.1.1 *Assessing the Rooftop Solar Potential — Case Study of Mumbai*

A preliminary study on the load profile of the city of Mumbai and the expected generation profile from PV installations was carried out recently [Singh and Banerjee (2015)]. The study concluded that the total rooftop PV installation potential for Greater Mumbai is around 1.72 GWp. The residential buildings have a major share of nearly 1.3 GWp followed by industries (223 MWp), educational amenities (72 MWp), commercial buildings (56 MWp), office buildings (20 MWp) and hospitals (15 MWp). The railways can produce 25.5 MWp power from the rooftops available from their platforms and associated buildings in the stations within Municipal Corporation of Greater Mumbai (MCGM) boundaries. The 25 bus depots across MCGM can have another 4.4 MWp on their rooftops. We were also able to identify some of the major structures under each category which have a potential of more than 100 kWp. Further technical studies need to be done from a grid integration point of view of rooftop systems from such huge available shade-free areas across the city.

The results show that even though the entire 1.72 GWp potential is realised in installations, the grid stability may not be affected in terms of real time load management. In fact, considerable PV penetration into the grid can help the utilities in managing the morning peak demand and ramp rates. Given the potential sources, a study on the points of sinks (major load centres of electrical power) can make the picture clear about how effectively the rooftops can be utilised. If there can be any matching between the load centres and the rooftop availability across various buildings, then it can lead to effective designing of rooftop PV systems feeding the needs of nearest load centres. Such a study needs the support of the utility companies.

6.1.2 *Building Integrated Photovoltaics (BIPV)*

BIPVs are becoming economically viable as a building material due to the decline in their cost. In urban areas, the potential to harvest solar energy by facade is larger compared to the roof. Its potential depends not only on the geographical location but also on surrounding entities. Since

BIPV has to be implemented in the design phase of the building, accessing its potential for deploying appropriate technology early in the life cycle is important. From the building profile, we need to first identify the solar potential for the building to decide whether to have BIPV or not. In urban areas, the solar potential is greatly affected by the shadows falling from the surrounding objects such as trees, buildings, billboards, etc. Solar potential is also influenced by geographical parameters such as climate, longitude, and latitude.

Another way of estimating the potential is to apply an empirical model on the map through the Geographical Information System (GIS) tools. These models use statistical analysis and regression models [Bose (2010)] to evaluate the potential. The work [Lariviere and Lafrance (1999)] discusses city-scale roof potential estimation from both 3-D data as well as 2-D to estimate the rooftop potential. However, for facades, the level of detail (LOD) 2.5 or 3 is required to assess the solar potential. The definitions of various LOD levels of building modeling and their implications can be found in [Biljecki *et al.* (2016); Kolbe (2009)]. A calculation method to estimate the facade potential using a 3D model of the building and surroundings is proposed in [Colson and Nehrir (2009)]. Since most cities lack 3-D details, it is cumbersome to create city-level or a neighborhood scale 3-D model and simulate. In India, every city has building codes or bye-laws especially fire safety norms which govern the building height and distance needs to be maintained around the surrounding entities (building, road). To overcome this limitation, a model has been developed [Jois *et al.* (2020b)] by transforming the entities into spherical coordinates. This transformation gives the shadow mask for the building. These findings are validated by placing the BIPV panels on a building and taking the measurement for one year.

Once we have the radiation profile of the building, we need to know which technology would be suitable for the building facade. The efficiency of BIPV varies depending on the solar radiation i.e., for direct, diffused, and reflected radiation. To estimate the yield of different BIPV technologies, we need to calculate and analyze the solar spectrum of the incident radiation. Spectrum response of any BIPV technology depends on the bandgap of the material. Solar geometry is used to estimate the yield. A prototype was developed to determine the yield as well as the human comfort impact of available technologies. Depending on the climate and orientation, BIPV technologies are chosen such that it will minimize the electricity consumption for air-conditioning and also generate electricity.

Heat transfer through the building envelope gets affected by BIPV. This is because, installation of BIPV alters the thermal resistance of the envelope by adding or replacing the outdoor structural/surface elements of buildings. Roofs with different kinds of BIPV installations have been studied in [Wang *et al.* (2006)]. Simulation studies have been reported on the impact of BIPV on the thermal environment of buildings. A study in [Yang and Athienitis (2014)] considered roof based BIPV and another one in [Ekoe a Akata *et al.* (2015)] considered the impact of facade based BIPV.

Very few studies compare different technologies and transparencies. The authors of [Lopez and Sangiorgi (2014)] compared Copper Indium Selenide (CIS), Amorphous Si (a-Si), and Monocrystalline technologies. None of the studies considers different technologies and transparencies at the same location to compare the performances. [Aelenei *et al.* (2014)] studied different heat gain mechanism for facade based solar. As the impact of the BIPV varies w.r.t. transparency and technology, a study comparing multiple technologies of single transparency on electricity generation and their impact on indoor thermal comfort remains a lacuna. A study on the thermal impact of BIPV on the indoor environment has been carried out in [Jois *et al.* (2020a)]. In this study, the impact of six BIPV technologies, and four different transparencies with four orientations on the indoor thermal gain is investigated. Two solar PV structures consisting of BIPV modules of six different technologies a-Si, CdTe, c-Si, CIGS, Micromorph, and Triple junction (Si-Ge) and four different transparencies were considered. Each technology has varying transparencies — opaque to semi-transparency of up to 40%. Results obtained on a summer day show that

- The thermal gain of the triple junction was lowest and electricity generation was higher for monocrystalline.
- 30% transparency modules have lower thermal gain and higher electricity generation compared to opaque, 10% and 20% (except the CIGS technology).
- The current location's South-West, South-East, and North-West orientations offer less thermal gain and more power generation compared to North-East orientation.

Hence, it would be viable to implement certain facade based BIPV technologies and transparencies depending on the geography, orientation, and climatic condition for efficient use of renewable energy. Selection and deployment of these technologies will be guided by the computed BIPV yield for the specific applications. For Mumbai, the triple junction thin-film PV module on flexible substrate is a viable option on facades (for its lightweight, mechanical flexibility and better sunlight to electrical energy efficiency) to reduce the thermal heat gain.

6.1.2.1   *BIPV Optimization*

In order to maximize the yield, the BPIV modules must be placed optimally along with their optimally configured electrical connections. There are different methods of deploying BIPV. They are integrated with part of the building, retrofitted on an existing structure or installed on the building skin. To achieve a good yield, they need to be mechanically fitted optimally.

Geometrical Programming [Boyd *et al.* (2007)] is one of the ways of framing the optimization problem which could be applied in practice. For example, floor planning is an optimization formulation that is widely used in the construction of building structures to VLSI layout design. It is solved using geometrical programming. Module placement optimization is similar to floor planning. The objective in both cases is to fit the given area with maximum basic building blocks. The basic building block, in this case, is a single BIPV module. But optimal placement of BIPV is more complex than floor planning or VLSI circuit gate placements. This is because, in addition to area, solar insolation profile is an additional constraint to be considered. In other words, the effect of shading/partial-shading caused by the surrounding structures introduces an additional constraint in the optimization problem of maximizing the BPIV yield. In [Jois *et al.* (2020b)], a building wall is modeled as a matrix of grids with different insolation level and the optimization problem is solved for a small grid. The solution indicates that instead of a linear array or cascading, tilt in the position of the vertical or horizontal end can maximize the yield.

Next, the electrical assembly has to be optimized so that electricity generated is collected with minimal loss. For a given building, we need to know how many BIPV modules should be connected in series and parallel combinations and how many power converters would be required.

6.1.2.2   *Partial Shading and MPPT*

Maximum power point tracking (MPPT) is essential to maximise the output of any solar panel [Sundareswaran *et al.* (2015); Seyedmahmoudian *et al.* (2019)]. However, due to the non-availability of direct radiation throughout the day, most of the time the PV modules will be subjected to partial shading conditions. Partial shading in a PV array, in some cases, can create a short-circuit like condition and thus result in damage to the PV modules due to over-current [Karmakar and Pradhan (2020)]. In order to protect the modules, bypass diodes are connected across the modules. However, in the presence of bypass diodes, multiple peaks are observed in the power versus voltage characteristic of the PV array during partial shading. In other words, the BIPV array will have many local maxima and therefore, application of superior MPPT algorithm is required to derive the maximum power output. MPPT algorithm is a well researched area. Apart from the popular perturb-and-observe (P&O) algorithm, many sophisticated algorithms including a number of AI based variants are available in the literature [Sundareswaran *et al.* (2015); Rizzo and Scelba (2015); Lian *et al.* (2014)]. However, MPPT tracking alone cannot exploit the full power generation potential of a PV array under partial shading.

6.2 Mitigation of the Effect of Partial Shading

6.2.1 *Output Control using MPPT*

Modern solar-powered systems already actively control solar power output within their inverter, which converts the DC electricity generated by the solar modules into AC electricity, which is synchronized with the grid's frequency and phase. Inverters typically implement an embedded algorithm for Maximum Power Point Tracking (MPPT) that constantly adjusts the deployment's operating voltage to maximize its power generation, as the current produced by solar modules varies nonlinearly with voltage.

The primary factor that affects a solar deployment's maximum possible production is its solar insolation, i.e., the amount of solar radiation that is incident on the solar modules' area. The amount of solar insolation is affected by numerous variables, including the weather, angle of the sun in the sky (which varies across the day and year), shade from neighboring buildings and trees, modules' tilt and orientation, etc. Given these factors, a typical solar module is capable of operating at a range of different current and voltage levels, which govern its actual power output. The operating region of a solar system is governed by its I-V curve, as depicted in Figure 6.1. The figure shows a solar module's output current across a range of voltages (as dictated by the applied resistance), where the solar power output is simply the product of the voltage and current. Due to the nature of the I-V curve, the solar output power changes at different operating

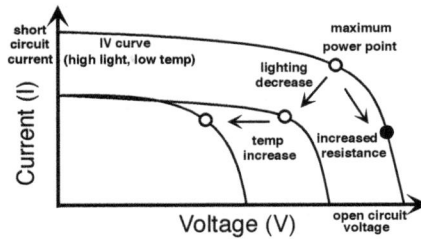

Fig. 6.1   I-V curve for a typical solar module, and the effect of changes in lighting and temperature.

voltages. Specifically, since the I-V curve is initially flat, as the operating voltage increases, the output current remains virtually unchanged, leading to an increase in power output. However, after reaching the knee of the curve, any further increase in operating voltage yields a corresponding reduction in current, and hence the power begins to drop. Thus, the solar output rises with increasing voltage up to a point and then precipitously drops. As a result, each I-V curve has an optimal operating voltage V_{opt} that maximizes its output.

Note that the precise shape of the I-V curve is dynamic and changes continuously. For example, the maximum power point decreases as the solar insolation decreases, causing the curve to contract along both the x-axis and y-axis as depicted. In addition, the solar cell temperature also affects the shape of the curve, expanding and contracting it along the x-axis. While Figure 6.1 depicts an idealized curve for a single solar module, solar systems are typically composed of multiple modules wired (or "strung") together and connected to a single inverter. In this case, the I-V curve of the aggregate solar circuit is a combination of the I-V curves of each module. Figure 6.2 shows how the combined I-V curve is a composition of each module's I-V curve when wiring modules in series (Figure 6.2(a)), in parallel (Figure 6.2(b)), and a combination of the two (Figure 6.2(c)). In particular, two modules wired in series operate at the same current, but have additive voltage, while two modules wired in parallel operate at the same voltage, but have additive current. The characteristics of each module may then change independently, affecting both the output of the other modules and system's aggregate I-V curve. For example, two connected modules may be installed with different tilts at different orientations, causing a shadow to cover one but not the other. If wired in series, the module producing the lowest current restricts the current generated by the other modules, reducing the entire array's output.

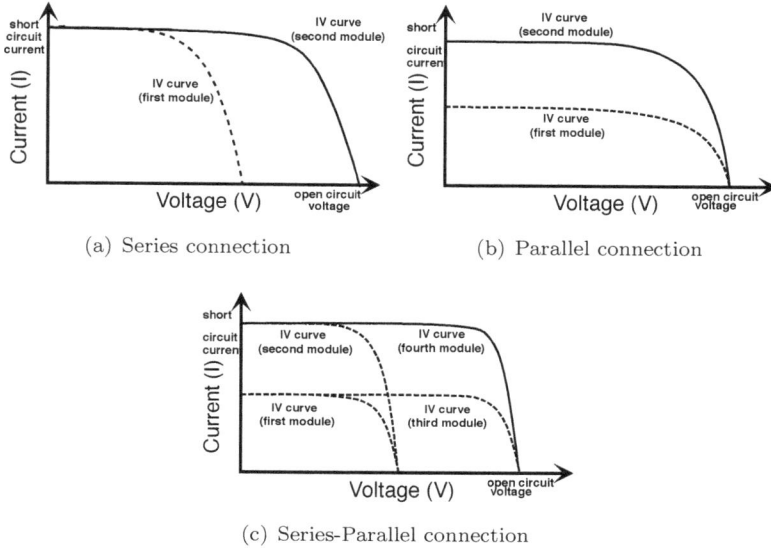

(a) Series connection

(b) Parallel connection

(c) Series-Parallel connection

Fig. 6.2   Example of I-V curves for a solar array formed from wiring multiple solar modules.

MPPT algorithms dynamically adjust the system's voltage to maximize power generation by operating at the "knee" of the I-V curve as the curve changes. Inverters implement MPPT algorithms using a DC-to-DC buck-boost converter that is able to adjust output voltage to be greater than or less than input voltage. Buck-boost converters typically use pulse-width modulation (PWM) to vary their duty cycle, which also varies the input/output voltage. There is a large body of work on developing Maximum Power Point Tracking (MPPT) algorithms — examples include the perturb and (P&O) observe, current sweep, incremental conductance, and constant voltage ratio algorithms amongst many others. MPPT algorithm design is well-studied and presents many trade-offs in optimizing power accuracy, convergence speed, implementation complexity, initialization procedures, etc. [2].

The most common MPPT algorithm is the Perturb and Observe (P&O) algorithm. This algorithm perturbs the voltage by a small amount, and then measures the instantaneous current and voltage to calculate the new power (P_t) and compares it to the power P_{t-1} at the previous voltage. If the change in power is positive, it continues to perturb the voltage in the same direction; if the change is negative then it reverses the direction of its search. Simple P&O algorithms use a fixed voltage step size on each iteration, while more sophisticated variations adapt the step size, e.g., proportional to the slope of the P-V curve $\frac{\Delta P}{\Delta V}$, to converge more quickly.

6.2.2   *Beyond MPPT Control*

It has been observed that the full generation potential of a PV array during partial shading cannot be derived by using MPPT controllers, irrespective of their level of sophistication.

In order to increase the array output power beyond what is achievable using MPPT controller alone, techniques like i) PV array reconfiguration [Velasco-Quesada *et al.* (2009); Salameh and Dagher (1990)] and ii) current injection using additional converters [Busquets-Monge *et al.* (2008); Sharma and Agarwal (2014)] are proposed. However, a hybrid approach [Karmakar and Karmakar (2021)] to derive the benefits of both dynamic array reconfiguration (DAR) and current injection (CI) can offer the full power generation capacity (under partial shading) of a PV array. Also, this hybrid CI-DAR approach requires less number of converters than what is required by CI approach alone. Note that none of these DAR, CI and CI-DAR techniques replace MPPT, rather they enhance the output power while MPPT is in place.

> DAR, CI or CI-DAR techniques do not replace MPPT. Rather, they enhance the output power while MPPT is in place.

We will describe the above approaches and compare them with the example of a 5×3 total cross-tie (TCT) connected solar PV array as shown in Figure 6.3.

6.2.3   *Dynamic Array Reconfiguration (DAR) and Current Injection (CI)*

In a solar PV system, *dynamic array reconfiguration (DAR)* of the electrical connections among the PV modules can improve the output in the event of partial shading. In this approach, the connections of partially shaded PV modules are electrically rearranged (using relay-based switches) so that the modules are equally distributed among the rows to the extent feasible. Consider the example shown in Figure 6.3(a), where two PV modules

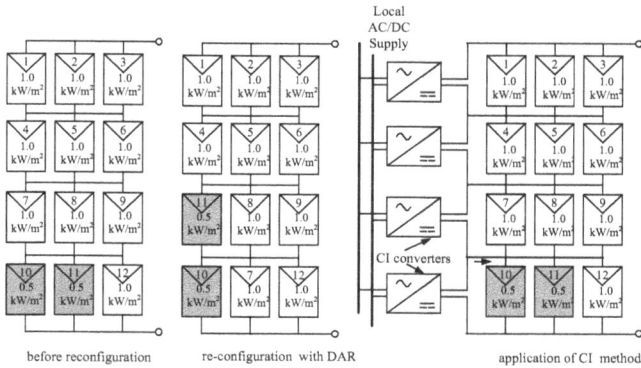

(a) Case 1: CI wins over DAR.

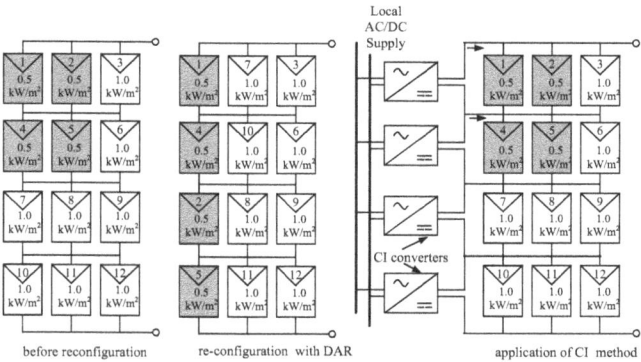

(b) Case 2: CI and DAR both are equally effective. But DAR is preferred for lower investment cost.

Fig. 6.3   Comparison between DAR and CI.

(modules #10 and #11) are shaded in the bottom row. It can be observed from the figure that before reconfiguration, the array current is limited to the current generated by the bottom row with shaded modules. However, if the electrical positions of two shaded modules are rearranged as shown in the middle image of Figure 6.3(a), it will enhance the array current.

Alternatively, as it can be observed from the rightmost image of Figure 6.3(a) that if we inject enough current (by using external converter) in the bottom row enabling them to operate at their MPPT, it will counter the effect of the shading in the modules. This is termed as the *current injection or CI* technique.

It can further be observed from Figure 6.3(a) that though DAR improves the array current, it is still limited by the lowest current produced by the rows containing shaded modules. In contrast, when CI technique is applied, it helps extract the full power generating potential of the PV array. But, it must be noted that this does not imply that CI is superior to DAR. Their advantage over the other depends on the number of shaded modules and the shading pattern. Let us explain this with the help of Figure 6.3(b) with four numbers of shaded modules (module #1, #2, #4, #5). It can be observed from the figure that by equally distributing the shaded modules in the rows, DAR ensures same power production capability by each row. The same is achieved by using CI techniques also (refer to the rightmost image of Figure 6.3(b)). Therefore, theoretically, both these techniques are equally effective in this case. However, it can also be observed from the figure that implementation of CI requires deployment of one converter in each row and in this case, only two converters (at the top two rows) are utilized. Therefore, DAR is preferable over CI here because, DAR involves much lower investment, which requires only switches.

The above discussion brings out the following facts.

- Achieving the full generation potential of a PV array is not guaranteed by DAR.
- In a real-world scenario of partial shading, deployment of one converter in each row is often under-utilised.

6.2.4   CI-based DAR (CI-DAR)

The hybrid technique of CI-based DAR brings out the best of both CI and DAR approaches. This approach has the following advantages [Karmakar and Karmakar (2021)].

- Performs better than DAR in mitigating the effect of partial shading on the PV array output power.
- Prevents multiple peaks in the current-voltage (or, power-voltage) characteristics of PV array.
- Can be more economical compared to CI alone as it uses less number of converters.

In this approach, first, rearrangement of the electrical connections of the shaded modules is carried out dynamically (as in DAR) to bring down the difference in current production in the rows to minimum — preferably to zero. However, the difference in current production can be brought down to zero by DAR, only if the number of shaded modules s in an $m \times n$ array

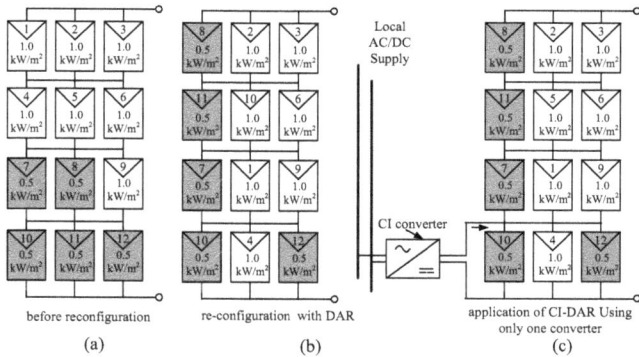

Fig. 6.4 Mitigation of partial shading with CI-DAR using minimum number of converters under uniform partial shading. [Karmakar and Karmakar (2021)]

are such that s is an integer multiple of the number of rows ($s = \mathbb{Z}^+ \times m$). If s is not an integer multiple of m, then attempt is made to minimise the difference in the row currents to the extent that it is possible by rearranging the shaded modules as shown in Figure 6.4(b).

But, it may be possible to enhance the array power output further, if a rearrangement is made and current is injected in the relevant row(s) using switchable converters as shown in Figure 6.4(c). It can also be observed from the figure that by applying this hybrid approach of *CI followed by DAR* the full generation potential of the array can be exploited with lesser number of converters than what would have been required by conventional CI alone. This technique is termed as CI-DAR.

In real-world applications, only partial shading, i.e., shading of only few PV modules in an $m \times n$ array is observed and therefore, it makes sense to deploy fewer than m number of converters based on historical data of partial shading.

It may be noted here that the power supply to these switchable converters can be provided from the local AC/DC bus or from the PV array itself. With the help of these converters, array power can be enhanced mitigating the effect of partial shading substantially.

6.2.4.1 *PV Array Size and Cost-Effectiveness of CI-DAR*

It has already been discussed in Section 6.2.3 that CI and DAR require additional investment in converters and switches, respectively. With the increase in array size, the requirement of the number of converters increases

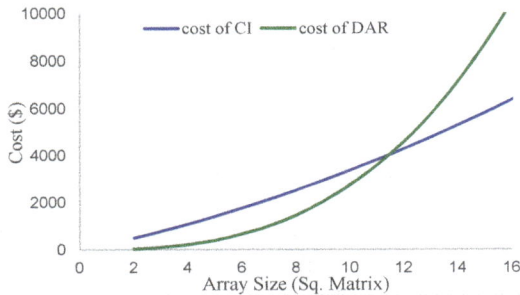

Fig. 6.5   Implementation cost of CI and DAR.

linearly. In contrast, the requirement of the number of switches increases exponentially with the increase in array size. CI-DAR involves both converters and switches and a study [Karmakar and Karmakar (2021)] shows that CI-DAR is more economical when the array size is not larger than about 12 × 12 as depicted in Figure 6.5. Since BIPV is prone to partial shading and mostly the size of an array in such applications is small, CI-DAR can be considered as the most suitable technique for improving power output in BIPV under partial shading.

6.2.4.2   *Current Injection (CI) along with Reconfiguration*

First, the pattern and the level of partial shading are estimated by measuring the voltage and current produced in each PV module as in [Velasco-Quesada *et al.* (2009)]. This is followed by the application of a reconfiguration algorithm to disperse the effect of shading uniformly throughout the array. If reconfiguration can achieve this partially and only for a limited number of rows, current is injected into the rows where relatively higher difference in row currents are observed. We use the term CI converters for those, which are used for current injection into the rows.

6.2.4.3   *CI-DAR Strategy*

Partial shading can be of two types — uniform and non-uniform. The strategies for mitigating these two types of shading and the corresponding requirement of the number of CI converters are different. What remains common is the first step of identifying the number of shaded modules s and their level of insolation.

Strategy for Uniform Partial Shading

If the number of shaded modules s happen to be an integer multiple of m, the number of rows in an $m \times n$ PV array, then application of DAR itself can help to extract maximum power from the array. If otherwise, the following strategies are applied.

Case 1 ($s < m$): Carry out reconfiguration by populating the rows(s) with shaded modules starting from the bottom (or top). This enables us to fill the number of row(s) with the shaded modules equal to $\lceil \frac{s}{n} \rceil$ with one row partially filled. Once this is done, switch on the CI converters connected to the rows containing shaded modules in order to compensate for the reduction of current in the PV modules caused by shading.

Case 2 ($s > m$): In this case, reconfigure the PV modules by populating the columns with shaded modules starting from the left (or right) so that entire column(s) is(are) filled with shaded modules only. Next, starting from the bottom (or top) row, replace the unshaded modules ($< m$) with the remaining shaded modules till all of them are exhausted. Switch on the CI converters to all the rows with unequal distribution of shaded modules.

Feasibility of CI-DAR — Number (minimum) of Converter Requirements:

In [Karmakar and Karmakar (2021)], it is shown that

The full power generation potential of a uniformly partially shaded $m \times n$ TCT connected PV array can be utilized by adopting the CI-DAR technique using N_c numbers of converters, if the following condition is satisfied.

$$\forall s < (m \times n), \quad N_c = \left\lceil \frac{\mathrm{mod}\left(\frac{s}{m}\right)}{n - \lfloor \frac{s}{m} \rfloor} \right\rceil \tag{6.1}$$

where, s is the number of shaded modules.

Strategy for Non-Uniform Partial Shading

In case of non-uniform partial shading, reconfiguration cannot guarantee uniform dispersion of the effect of partial shading.

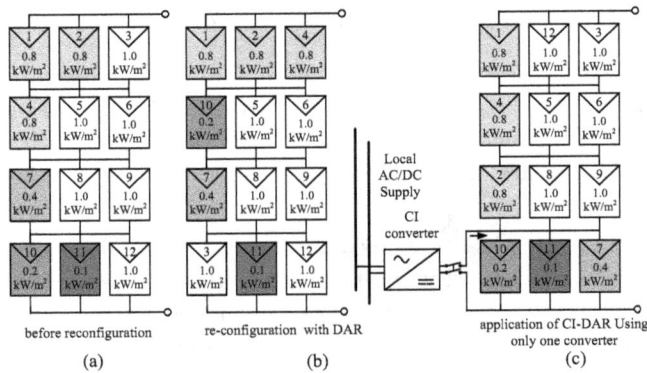

Fig. 6.6 Mitigation of partial shading with CI-DAR using minimum number of converters for non-uniform partial shading. [Karmakar and Karmakar (2021)]

A reconfiguration Algorithm 4 is used considering the availability of N_c number of converters according to (6.1). This algorithm works towards minimizing the differences in the row currents (caused by the shaded modules) considering all the possibilities to arrive at the best possible configuration.

In a real-world scenario, it is likely that some of the shaded PV modules will receive equal/nearly-equal insolation and can be considered as uniformly shaded. Our initial consideration of the availability of N_c number of converters will be sufficient in mitigating the effect of partial shading fully if

  i) the modules having large difference in the shading levels can be accommodated in N_c number of rows and
 ii) the remaining uniformly shaded modules can be equally distributed in the remaining $(m - N_c)$ rows.

However, if the Algorithm 4 concludes that uniform rearrangement of shaded modules is not possible within the $(m - N_c)$ rows, then more number of converters are required. However, the required maximum number of converters can be derived using the historical data of partial shading (offering finite and different levels of non-uniformity) utilizing the same Algorithm 4.

In order to appreciate the above idea, let us consider the example of a 4×3 array, shown in Figure 6.6 where the PV modules are subjected to non-uniform partial shading. The level of insolations in individual modules can also be observed in the figure. Figure 6.6(b) shows that conventional reconfiguration (e.g., using [Velasco-Quesada *et al.* (2009)] or [Babu *et al.*

(2018)]) cannot exploit the maximum power generation capacity of the array. However, it can be observed from Figure 6.6(c) that by applying Algorithm 4, maximum power generation potential of the array can be exploited when PV modules experiencing the same level of insolation are uniformly distributed in the top three rows of the first column and the rest are placed in the bottom row supported by one converter only.

6.2.4.4 *Mitigation of Non-uniform Partial Shading and CI-DAR Algorithm*

A configuration with minimum mismatch in insolations between any two rows provides maximum output power under partial shading [Velasco-Quesada *et al.* (2009)]. This criterion is valid and independent of whether CI converter is used or not.

However, under partially shaded conditions, the current flowing through a particular module may not indicate its actual current producing capability. This is because, the current in a module can be limited by some other shaded module (connected) in series. Thus a PV array is reconfigured based on the insolation received by the modules.

Before physical reconfiguration, the best possible reconfiguration is computed for a given number of CI converters N_c using a CI-DAR Algorithm 4. For this purpose, the algorithm needs the maximum number of feasible configurations C of the PV modules (in an $m \times n$ array) as an input, which is computed using the following equation [Velasco-Quesada *et al.* (2009)].

$$C = \frac{(m \cdot n)!}{m! \cdot (n!)^m} \tag{6.2}$$

The CI-DAR Algorithm

The CI-DAR algorithm [Karmakar and Karmakar (2021)] first computes the current producing capabilities of each row from the estimated insolation levels on the PV modules derived from their voltage and current measurements. Then it computes the best reconfiguration, which offers minimum difference in currents between any two rows.

Once the best configuration is available, maximum array power can be extracted with the application of the CI-based technique on it. This algorithm generates the optimum configuration because, brute force of checking all possible configurations is used. Note that this algorithm does not require high computing power for any practical $m \times n$ TCT PV array amenable to reconfiguration.

Algorithm 4 Algorithm to compute best possible configuration of an ($m \times n$) PV array using a given N_c no. of converters

---

procedure GETBESTCONFIGURATION(m, n, N_c)

    for $i = 1$ to m rows **do**

        Compute $I_r[i]$ ▷ The current producing capabilities I_r of each row, derived from estimated insolations.

    end for

    for $k = 1$ to C **do**   ▷ C – the number of possible configurations

        Initialize $\Delta I_m[k] = 0$ ▷ the maximum difference in currents between any two rows for k^{th} configuration

        sort $I_r[i]$ in descending order.

        find $\leq N_c$ number of rows, which produce lowest I_r among m rows and connect CI converters to these rows.

        for $i = 1$ to $(m - N_c)$ **do**

            for $j = (i + 1)$ to $(m - N_c)$ **do**

                Compute $\Delta I = |(I_{r(i)} - I_{r(j)})|$ such that i) $i, j \notin R_c \mid R_c = \{$converter connected rows$\}$ and ii) $j \neq i$

                if $\Delta I \geq \Delta I_m[k]$ **then**

                    $\Delta I_m[k] = \Delta I$ ▷ Store maximum difference in currents between any two rows for the kth configuration

                end if

            end for

        end for

    end for

    Select the kth configuration with $\min(\Delta I_m[k])$.   ▷ Offers best configuration with minimum possible difference in currents between any two rows, thereby maximizing array power output with CI-based reconfiguration

    return k;

end procedure

Note that *Physical Reconfiguration* is carried out by switching the PV modules to arrive at the best possible configuration computed. In this context, it may be noted that in order to avoid the partial shading created by quick passing clouds, a delay of about 20 s should be introduced before switching to a new configuration.

6.2.5 *Experimental Validation in a Prototype System*

An experimental validation in a prototype system using PV modules, DC-DC boost converters and CI converter was carried out [Karmakar and Karmakar (2021)] to evaluate the performance of CI-DAR.

6.2.5.1 *The Prototype System*

An experimental setup of 3×3 TCT configured PV array was developed using commercial solar modules, over-the-shelf electronic hardware and CI converters. The specification of the PV modules used in this experiment is shown in Table 6.1. A Sun simulator [Eternal Sun] is used as the source of solar irradiance on the PV modules and partial shading is created with the help of a ink (black) spotted piece of white cloth. For this experiment, a resistive load is connected to the PV array through a PWM controlled DC-DC boost converter. Conventional perturb and observe (P&O) MPPT algorithm is applied to achieve maximum power output from the PV array. An AC-DC converter with a specification — *Input:* 230 V, 50 Hz, *Output:* 0–20 V, 0–2 A, is used as a CI converter, which receives power from local AC supply.

Table 6.1 Specifications of PV panel (Source: Manufacturer's data sheet).

Parameters	Value
Open circuit Voltage (V_{oc})	21.0 V
Short circuit Current (I_{sc})	0.61 A
Voltage, at MPP (V_m)	17.3 V
Current, at MPP (I_m)	0.57 A
Maximum Power (P_m)	10 W

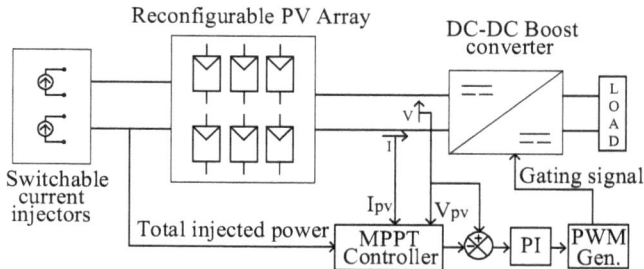

Fig. 6.7 The CI-DAR Experiment: Block diagram of hardware interconnections.

MPPT controller is simulated using an OPAL-RT simulator (model: OP–4510) and its output is fed to the PWM controlled DC-DC boost converter. This simulator also controls the CI converter used in this experiment. The overall scheme of this hardware experiment can be seen in Figure 6.7. The array current and voltage, injected current and the voltage across the CI converter are measured and they are used as well as stored in the OPAL-RT hardware in loop (HIL) simulator, which can be identified in Figure 6.8 — a photograph of the complete setup.

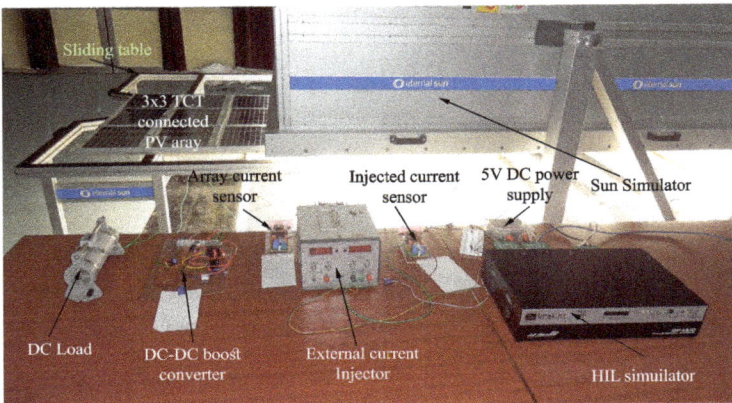

Fig. 6.8   The CI-DAR experiment setup. [Karmakar and Karmakar (2021)]

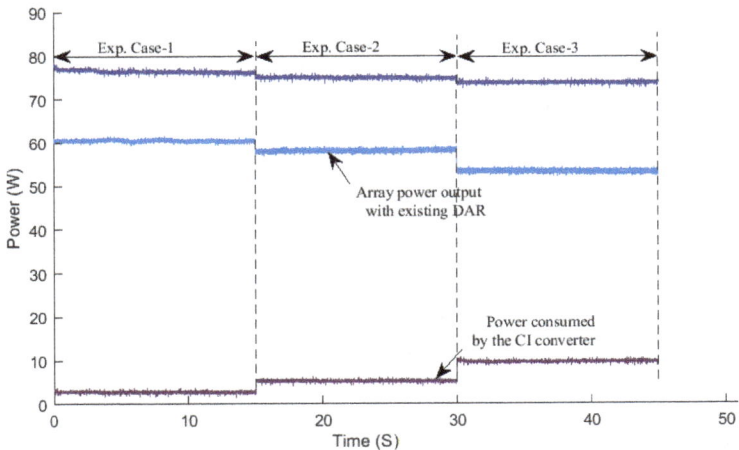

Fig. 6.9   Enhanced power output by CI-DAR.

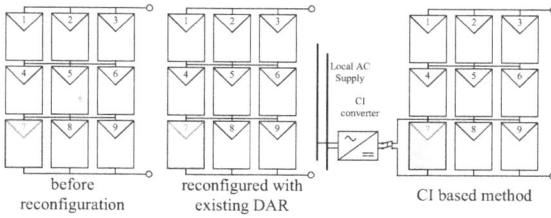

Fig. 6.10   Case 1: Conventional DAR is ineffective.

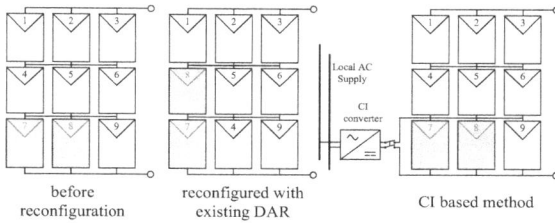

Fig. 6.11   Case 2: CI-DAR extracts more power compared to DAR.

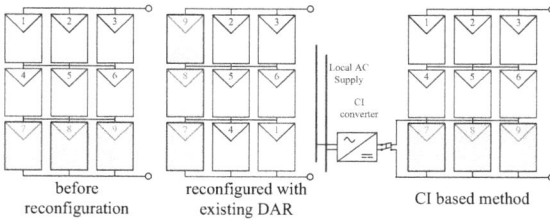

Fig. 6.12   Case 3: CI-DAR exploits full power generation potential of the array.

6.2.5.2   *The Experiment*

Three representative cases of partial shading studied under this experiment
are as follows.

Case 1 — One module shaded in the bottom row: This simple test configuration represents a case where, DAR is ineffective.

Case 2 — Two modules shaded in the bottom row: This configuration is
taken up to elaborate the case, where array power output can be improved
by DAR, but CI-DAR helps to improve the output further.

Case 3 — Three modules shaded in the bottom row: This case explains how

DAR can only disperse the effect of partial shading uniformly, but the full power generation potential can be exploited by CI-DAR only. An insolation level of 1.0 kW/m^2 produced by the Sun simulator is used for conducting the experiments pertaining to various representative cases of partial shading. An irradiance of 1.0 kW/m^2 above the cloth creates 0.40 kW/m^2 and 0.66 kW/m^2 insolation under the black spots and the white regions, respectively.

6.2.5.3 *Results*

The array power output data from the above experiment cases are plotted in Figure 6.9. It can be observed from the figure that at the expense of a small amount of power to the CI converter, the CI-DAR technique derives the full generation potential of the array for all three cases discussed above. The results are summarized numerically in Table 6.2.

Table 6.2   Experimental results comparing PV array power outputs.

Shading cases	With DAR (W)	With CI-DAR (W)	Power Consumed by CI converter (W)	Net gain in CI-DAR over DAR
Case-1	61.0	77.0	3.3	12.7 (20.8 %)
Case-2	58.5	75.0	6.7	9.8 (16.9 %)
Case-3	54.1	73.8	12.6	7.1 (13.1 %)

6.3 Summary and Takeaways

The need and the potential for solar energy in urban buildings are both enormous. But it comes with its own challenges. One of the important challenge is the varying insolation on the PV arrays, which is aggravated by partial shading. MMPT cannot take care of partial shading, which causes significant reduction in PV array power output. The CI-DAR technique of mitigating the effect of partial shading is discussed and compared with the conventional DAR and current injection (CI) approaches with the help of experimental results, which is encouraging.

Chapter 7

Making the Best of Available Energy

In this chapter, we consider two additional topics in energy management related to power deficit, i.e., when power generated is less than the power needed. During *power deficit* periods, utilities resort to scheduled blackouts in the sub-areas, which may adversely affect the consumer. On the other hand, based on the analysis of the data on users' power consumption pattern gathered through Non-Intrusive Load Monitoring (NILM), utilities attempt to educate and influence (often by penalizing) the consumer towards energy-aware usage of loads.

7.1 Managing Building Loads According to Available Power in the Grid

In many developing countries like India consumers experience *power deficit*. The conventional mechanism adapted by regional operators to address the problem is *Rolling Blackouts*. Rolling Blackout is the *engineered electricity shutdown* for *non-overlapping time intervals* over different *sub-regions or sub-areas* within the power distribution region. The major side effect of this blackout solution is the complete disruption of electricity supply even to the essential loads which may include hospitals, business critical establishments and other basic conveniences like lights, fans, etc. However, if the building loads are managed by following the available power in the grid and the demand is reduced accordingly by limiting consumption only for the essential loads, blackouts can be prevented. We term this grid following load management solution as grid following brownout (GFB).

7.1.1   *Dealing with Blackouts: Existing Approach*

The existing approach is leaving it to the consumers to solve the problem, which depends on their ability to make own arrangements during a blackout. At the commercial building or large housing society level, in-house generation using diesel generators is common. However, at the small building/apartment and small store level, people often use battery-backed inverters.

7.1.2   *GFB: A Smarter Solution to Prevent Blackouts*

A *grid following brownout* (GFB) based *electricity scheduling* strategy allows the utility to effectively distribute the available power among a set of subareas during power deficit periods. Brownout refers to controlled distribution of a limited amount of power during *demand overloads* such that uninterrupted power supply to essential appliances/subareas (higher priority) may be selectively maintained while cutting down supply to less critical loads (i.e., loads with lower priority). The same philosophy can be extended to the consumers at the residential building level so that loads in individual buildings can be reduced by giving priority to essential loads like fans, lights and refrigerators. With Smart Grids and Smart Meters, the design of such improved schemes which essentially depend on more precise knowledge and finer controlability over consumer loads on the demand side, have now become practically realizable. Figure 7.1 depicts the GFB scheme vis-á-vis rolling blackouts. It can be observed from the figure that the brownout scheme can prevent blackout of the entire electrical loads connected to Feeder F1.

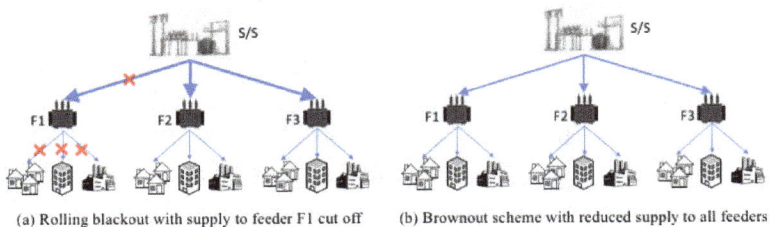

(a) Rolling blackout with supply to feeder F1 cut off   (b) Brownout scheme with reduced supply to all feeders

Fig. 7.1   Grid following brownouts (GFB) versus rolling blackouts.

> **Brownout**
>
> Brownout refers to a controlled distribution of a limited amount of power during *demand overloads* such that uninterrupted power supply to essential appliances or subareas (containing establishments like hospitals) with high priority electrical loads may be selectively maintained while cutting down supply to less critical loads with lower priority.

The GFB solution can be obtained by offering a technological solution to the following problem.

Given a system-wide upper-bound on consumption in times of power scarcity, how do we ensure that the available power is equitably distributed to the consumers, taking into account the criticality of their needs?

7.1.2.1 *A Comprehensive Framework for GFB-based Electricity Scheduling*

We now describe a framework that is essentially inspired from models proposed for *resource allocation strategies* typically used for Quality of Service (QoS) oriented scheduling mechanisms for *real-time* applications. The framework is also guided by the following *soft real-time* specification governed by recommendations of IEEE standards on power quality [IEEE Std 141; IEEE Std 1159]. The constraint in brownout based mitigation of imbalances in demand-supply is that it must be executed quickly and within about ~0.5 seconds (for a 50 Hz power system) in order to maintain satisfactory power quality.

Optimization Formulation of Brownout

The system is modeled as a set of N subareas $N = \{S_1, S_2, \cdots, S_N\}$ under the zone of distribution of a power utility. Each subarea consists of a collection of establishments, commercial and residential buildings. Electrical loads (appliances) within the establishments or the establishments themselves are then partitioned into a set of distinct equivalence partition classes, where the priority of the loads (or establishments) is determined by the criticality level of the requirements of uninterrupted power supply. For example, lights and fans in households and the establishments like hospitals can be considered as *critical* or *essential* and therefore allocated highest priority.

The remaining loads (like ACs, dishwashers, washing machines, etc.) in the subareas are considered to belong to the broad category of non-essential loads. The utility provides K_i priority levels $\{l_i^1, l_i^2, \cdots, l_i^{K_i}\}$ of the non-essential loads within each subarea S_i with distinct price rates based on their priority. It is a consumer's choice to assign the desired priority levels to their non-essential loads. Thus, the loads in a subarea S_i essentially gets divided into $(K_i + 1)$ disjoint subgroups, of which the priority level l_i^0 corresponds to essential load.

Let P_i^0 and r_i^0 denote the power demand corresponding to highest priority l_i^0 of S_i and rewards earned by satisfying P_i^0, respectively. Let P_i^j represent the power demand of all the non-essential loads contained in priority level 1 through j. After satisfying the total power demand of critical/essential loads, a revenue aware allocation/distribution of the residual power P_R is offered among the subareas so that the reward of the utility is maximized as in (7.2). The residual power P_R is formulated as

$$P_R = P_G - \sum_{i=1}^{N} P_i^0 \tag{7.1}$$

where, P_G is the total power available with the utility for distribution.

$$Maximize \sum_{i=1}^{N} \sum_{j=1}^{K_i} r_i^j \times x_i^j \tag{7.2}$$

$$Subject\ to \sum_{i=1}^{N} \sum_{j=1}^{K_i} p_i^j \times x_i^j \leq P_R \tag{7.3}$$

$$\forall i \in [1, N],\ x_i^j \in \{0, 1\},\quad \sum_{j=1}^{K_i} x_i^j \leq 1 \tag{7.4}$$

where, x_i^j is a binary variable, which is set to 1 if subarea S_i is the allocated power at priority level l_i^j. The first constraint guarantees that the total amount of power allocated to all the subareas remains within the total available residual power P_G. The second constraint forces each subarea to select at most one priority level. Equation (7.2), therefore is essentially a *Multiple Choice Knapsack Problem (MCKP)* [Basina *et al.* (2020a)], which is NP-hard.

Solution Strategies

Dynamic programming (DP) provides a natural solution for this optimization problem, which was studied in [Basina *et al.* (2020a)]. However, it was observed that the optimal solutions by Dynamic Programming (DP) when applied proves to be prohibitively expensive in terms of computational overheads (hence, violate soft real-time requirement). Therefore, the DP technique is modified and a new strategy is devised known as Streamlined DP-based Priority level Allocator (SDPA). SPDA capitalizes on the *discrete* nature of the power demands of subareas to work with a far lower number of non-dominating partial DP solutions and allows the ultimate optimal solution to be generated much faster. Even SDPA may fail to meet the real-time requirements of dynamic power imbalance mitigation in very large grids. Hence, a fast and efficient *heuristic* strategy namely, the proportionally balanced priority level allocator (PBPA), has been devised in [Basina *et al.* (2020b)], where it has been shown that the PBPA is approximately 10^4 times faster than SPDA.

7.1.2.2 *Consumers' Participation*

Brownout offers a limited amount of power to the consumer during a supply-demand mismatch scenario. Therefore, it requires participation of consumers to derive the best out of brownouts. Segregation of loads based on their priority (criticality) is a one-time task that must be carried out by individual users, which might require modifications in the existing electrical wiring in order to support brownout. On the other hand, utilities should be aware of the consumers' preferences or priority of the appliances so that effective brownouts can be planned accordingly. Static priorities can be assigned to different appliances (loads) along with a reward value for each of them so that utilities can reward or penalize the consumer based on their energy consumption during the energy deficit periods. However, a user would always prefer flexibility of the dynamic reconfiguration of their preferences/priority of the appliances. A hybrid approach combining the benefits of both static priority assignments of appliances and considering consumer preferences optimally is presented in [Ramanujam *et al.* (2020)].

7.1.2.3 *Brownout Algorithm*

A brownout algorithm is proposed in [Ramanujam *et al.* (2020)] for controlling power consumption in households. The loads are assumed to be prioritised as critical/essential (E), high (H), medium (M) and low (L).

In the event of total power demand P_D exceeding the threshold

---

Algorithm 5 Brownout Algorithm

1: **procedure** GETAPPLIANCESTATE(P_D, P_{th})   ▷ P_D= Total demand and P_{th}= Power threshold
2:     Get attributes $A_i(i, r^i)$ of all appliances   ▷ Attributes of ith appliance (i = priority level, r^i = its associated reward value)
3:     **if** $P_D \leq P_{th}$ and there is no change in appliance attributes **then**
4:         Do Nothing
5:     **else**
6:         Find priority E appliances with *desired-state*=TRUE   ▷ priority E = essential
7:         $A_k(k = 0, 1, \cdots, n - 1)$ ▷ n = Number of essential (Priority E) appliances
8:         Calculate essential power $P_E = \sum_0^{n-1} P_k$ ▷ P_k = Power consumption of appliance A_k.
9:         Turn ON all the priority E appliances
10:         Calculate remaining power $P_R = P_{th} - P_E$
11:         Find list of all the priority H appliances with *desired-state*=TRUE   ▷ priority H = high
12:         Sort appliance list $A_j^i(j = 0, 1, 2, \cdots, m^i - 1)$ in descending order of r_j^i/P_j^i ▷ i= H (high priority level), r_j^i= reward value, P_j^i = consumption estimate of jth appliance with priority level i.
13:         **while** $j < m^i$ **do**
14:             **if** $P_R > 0$ **then**
15:                 Allocate power to A_j^i
16:                 $P_R = P_R - P_j^i$
17:             **else**
18:                 Turn OFF A_j^i
19:             **end if**
20:             $j = j + 1$
21:         **end while**
22:         Repeat steps 11–21 with $i = M$ ▷ For priority M (Medium) loads
23:         Repeat steps 11–21 with $i = L$ ▷ For priority L (low) loads
24:     **end if**
25: **end procedure**

---

consumption[1] P_{th} or consumer changing the priority attribute of appliance(s), this algorithm first allocates power to the critical/essential (highest priority) load (or a set of loads). This is followed by allocating power to the loads (appliances) with next level of priorities in steps — from H to M to L — constrained by power P_R that remains available during a particular time period of the day. For a set of appliances having the same priority, the algorithm uses a greedy approach to maximize the rewards computed as r_j^i/P_j^i, where r_j^i and P_j^i are the reward and the power consumption of the jth appliance, respectively.

The pseudo code of algorithm is presented in Algorithm 5. The algorithm outputs the states (ON/OFF) of the appliances.

---

[1]Threshold consumption can have a different value during different time slots of the day depending on the dynamics of power generation and demand.

7.2 NILM: Non-Intrusive Load Monitoring

Non-intrusive Load Monitoring (NILM) is a technique to disaggregate individual loads from the aggregate power readings of an entire building or home. NILM was first proposed by George W. Hart [Hart (1989, 1992b)] in 1986. The main insight in NILM is that individual electrical loads have unique power usage patterns that manifest themselves in the aggregate power usage of the building, as observed by a smart meter, when the loads turn on or off or operate continuously. Consequently, given the knowledge of power usage of individual loads, it is feasible to decompose the total power usage into the constituent loads.

A NILM-based approach to sensing power usage has many advantages. First, it eliminates the need to deploy hard sensors at each load and allows the power usage of various load to be tracked using a single sensor — a smart meter — of sufficiently high sensing resolution. Second, the approach is non-intrusive to consumers, since no sensor needs to be deployed inside the home and all sensing and inference are done using a smart meter that is deployed at the ingress of the home. This enables utility companies to infer details of the power usage in a non-intrusive manner and run programs to educate and influence consumers regarding energy conscious usage of appliances.

7.2.1 *The Goal of NILM*

The intention of this technology is to collect data for load research for evaluating energy conservation options such as

i) research and development towards advancement in the appliance technologies,
ii) educating and influencing consumer concerning energy conscious usage of appliances, etc.

In the end, such exercises help facilitate consumers' participation in energy management, which includes demand-response (D-R) control of loads by flattening peak demand, laying out plans for mutually acceptable brownouts, etc.

> ## Goal of Non-Intrusive Load Monitoring (NILM)
>
> Collecting data for evaluating energy conservation options, which include
>
> (1) research and development towards advancement in the appliance technologies to make them
>
>   - more energy-efficient, and
>   - smarter by facilitating consumer-utility interaction.
>
> (2) educating and influencing consumer concerning energy conscious usage of appliances.

7.2.1.1   *Supporting Utility-Consumer Interaction*

Utilities attempt to influence the consumers by variable tariff so that peak demand can be reduced by shifting the timing of usage of higher power consuming loads like dishwashers and washing machines. The willingness of consumers' participation in the overall goals of smart grid and smart building can be facilitated by the use of NILM. This is because, the load data, like voltage and current, can be acquired from the energy meter panel outside the residence and therefore *non-intrusive* as it does not involve installation of any additional equipment inside the customer's home space.

7.2.1.2   *Identification and Segregation of Loads*

NILM aims at identifying the number and the nature of the individual loads (appliances) by analyzing the voltage and current waveforms of the aggregate load measured at the energy meter interface. NILM monitors the total load and looks for the specific *signatures*, which provides the operating conditions (ON, OFF and *running*) of constituent loads. For example, a step increase in power consumption of 1.5 KW, in a residence having a heater of same wattage indicates that the heater is turned ON and a decrease in power of that characteristic size indicates the turning OFF of the heater. Further, refrigerators, air-conditioners (AC) consume both active and reactive power and therefore these appliances can be easily distinguished from the resistive loads like room-heaters, geysers, toasters, iron, etc.

However, segregation of loads from the composite waveform of voltage and current available at the energy meter points is hard and remains an open problem today after it was first proposed wayback in 1986. In the following sections, we limit our discussion to the fundamental techniques of NILM along with examples.

7.2.2 *NILM Techniques*

NILM techniques, based on their load detection methods, can be broadly classified into two groups — i) edge detection and ii) model-driven.

7.2.2.1 *Edge Detection Methods*

The original NILM technique by Hart [Hart (1989)] was based on edge detection, and edge detection methods remain popular even today. In this approach, it is assumed that when a load turns on, it results in an increase in the total power consumption. By detecting an edge (i.e., a step increase in power usage), the technique can determine when a load has been turned on. Also since different loads consume differnet amounts of power, they will result in edges of varying magnitude. Hence, the height of the detected edge can point to the load that has been turn on. This simple idea has been used extensively to design a range of NILM techniques over the years.

7.2.2.2 *Model-Driven Methods*

More recent NILM techniques are model-based — they either assume that loads exhibit a certain behavior and look for this pattern within the aggregate power usage, or they use machine learning techniques to discern the presence of individual loads within the aggregate power trace. The former case requires a model for each load that is present and we discuss modeling techniques in the next section that can be leveraged by such model-driven techniques. In contrast to such techniques, data-driven modeling approaches based on machine learning have also become popular recently. These approaches can be viewed as classification problems, where machine learning or deep learning is used to disaggregate loads. We refer the reader to the following survey for further details of machine learning approaches for NILM [Nalmpantis and Vrakas (2019)].

7.3 Modeling Residential Electrical Loads

Many NILM techniques have begun to use specific characteristics of the load, rather than assuming that the load exhibits simple on-off behavior.

To enable such NILM techniques, we present models to capture a range of residential electrical loads.

7.3.1　*Modeling Individual Loads*

Appliances can be classified into three broad categories based on their operational states and can be modeled accordingly for the purpose of NILM (Non-Intrusive Appliance Load Monitoring). These three classes of models are:

- ON-OFF
- Finite State Machines (FSM), and
- Continuously variable

ON-OFF model, which is a boolean function, can model appliances like a light bulb, water heater, toaster, including thermostat controlled air-conditioners/heat pumps, which can be either ON or OFF. However, the washing machine, dishwasher have no distinct but finite number of ON states. Finite State Machines (FSM) are suitable for modeling this type of load. Appliances like variable speed drive based air-conditioners, sewing machines with continuously variable loads are difficult to model as these devices can have infinite number of ON states. In addition, there are some fixed electrical loads which are permanently ON. These loads are also to be identified for NILM to work. Example of continuously ON loads are telephone sets, cable TV sets (set top box) and home/building security system.

Using empirical observations, one can develop models to capture key characteristics of each load type. We first present four basic model types — *on-off, on-off growth/decay, stable min-max,* and *random range* — to describe simple loads, and then use these models as building blocks to form compound *cyclic* and *composite* models that describe more complex loads. Ideal models describe i) *how much* real and reactive power a load uses when active, ii) *how long* a load is active, and iii) *when* a load is active. However, in many cases, users manually control loads, such that *when* a load is active and for *how long* are non-deterministic. For example, a user may run a microwave any time for either ten seconds or ten minutes. For these loads, we assume a random variable captures this non-determinism, and focus our efforts, instead, on modeling *how* each load behaves when active.

7.3.2 Basic Model Types

7.3.2.1 On-Off Model

An on-off model includes two states — an *on* state that draws some fixed power p_{active} and an *off* state that draws zero, or some minimal amount of power p_{off}. Conventional on-off loads (see Figure 7.2). Dimmable lights also conform to on-off models, although p_{active} depends on the dim level. In this case, a $N\%$ dim level yields a proportionate reduction in real power usage. In addition, while real power is a simple linear function of dim level, reactive power is a quadratic function that peaks at 50% dim level. Constructing an on-off model is simple — we can use regression to determine appropriate values p_{active} and p_{off}. In particular, we partition the time series of load power usage into two mutually exclusive time-series, with data for the on and off periods, to determine the best values of p_{active} and p_{off}.

7.3.2.2 On-Off Growth/Decay Model

An on-off growth/decay model is a variant of the on-off model that accounts for an initial power surge when a load starts, followed by a smooth increase or decrease in power usage over time. AC induction motors are the most common example of a load exhibiting this behavior, e.g., refrigerator, central A/C, vacuum. Resistive loads with high-power heating elements, such as the toaster or coffee maker, also conform to an on-off growth/decay

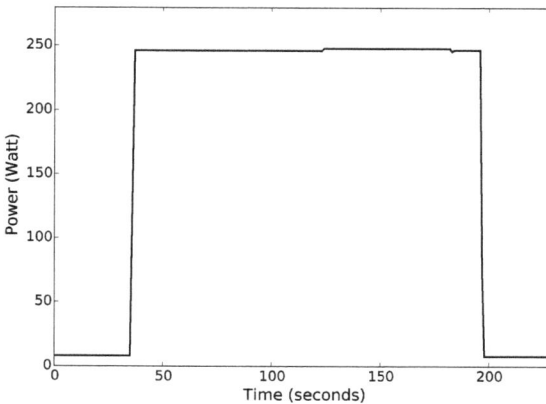

Fig. 7.2   Power usage of an incandescent light bulb follows the on-off model.

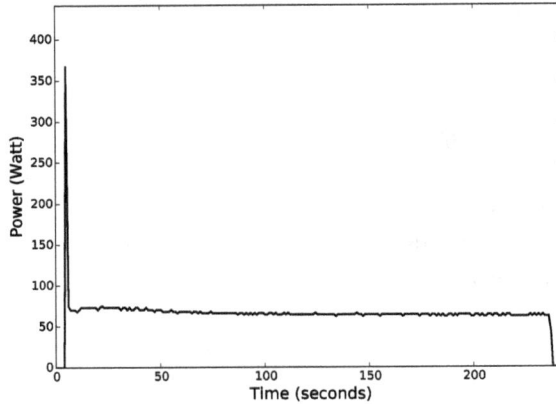

Fig. 7.3   A refrigerator exhibits an on-off decay model in its power usage.

model, although the surge and the decay in these devices are far less prominent than in AC motors. Figure 7.3 depicts how the compressor of a refrigerator exhibits an on-off decay behavior in its power usage. We can characterize on-off decay models using four parameters: p_{active}, p_{off}, p_{peak}, and λ. The first two parameters are the same as in on-off models, while p_{peak} represents the level of inrush current when a device starts up and λ represents the rate of growth or decay to the stable p_{active} power level. We model decay using an exponential function as follows, where t_{active} is the length of the active interval.

$$p(t) = \begin{cases} p_{active} + (p_{peak} - p_{active})e^{-\lambda t}, \ 0 \leq t < t_{active} \\ p_{off}, \ t \geq t_{active} \end{cases} \qquad (7.5)$$

Similarly, we can model on-off growth as a logarithmic function (i.e., the inverse of the exponential function) using starting power level p_{base} and growth parameter λ:

$$p(t) = \begin{cases} p_{base} + \lambda \ln t, \ 0 < t < t_{active} \\ p_{off}, \ t \geq t_{active} \end{cases} \qquad (7.6)$$

We can optionally augment the growth model with an additional parameter p_{ceil} to prevent unbounded growth that simply caps the maximum output of the model. In the growth model, the surge current must also be modeled separately. Here, we can simply add a parameter p_{spike} specifying the power at $t = 0$.

Constructing an on-off growth/decay model requires fitting an exponentially decaying (or logarithmically growing) function onto the time-series data, in addition to determining p_{peak}, p_{active}, and p_{off}. The LMA algorithm [Levenberg (1944)] can be used to numerically find the exponential or logarithmic function that best fits the data, i.e., based on a least-squares nonlinear fit.

7.3.2.3 *Stable Min-Max Model*

While on-off and on-off decay models accurately capture the behavior of resistive and inductive loads, they are inadequate for modeling nonlinear loads. Many nonlinear loads maintain a stable maximum or minimum power draw when active, but often vary randomly and frequently from this stable state. These variations are due to the device regulating their electricity usage at a fine grain to instantaneously "match" the needs of the tasks the device is performing. Our stable min-max model captures this behavior. We model stable min-max devices as having a stable maximum or minimum power denoted by p_{active} when active. The power usage then deviates, or "spikes," up or down from this stable value at some frequency. The magnitude of the spike is a random value uniformly distributed between p_{active} and p_{spike}, where p_{spike} denotes the maximum deviation in power per spike. The interarrival time of the spikes are exponentially distributed with mean λ. Thus, the stable min-max model has three parameters: a stable max or min power usage p_{active}, the maximum deviation p_{spike} for each spike, and λ, which governs the interarrival time of spikes. Note that p_{active} may denote either the stable maximum or minimum power, with power deviations decreasing or increasing, respectively.

Empirically constructing a load-specific stable min-max model requires determining the stable power level p_{active} and characterizing the magnitude and frequency of the power spikes. We employ simple regression to determine the stable power level p_{active} from the data, e.g., after filtering out the data for spikes and finding the fit for p_{active}. The mean observed duration between spikes then yields the parameter λ. Figure 7.4 shows a LCD TV that exhibits a stable-max behavior for its power usage.

7.3.2.4 *Random Range*

Some devices draw a seemingly random amount of power within a fixed range when active. This is likely due to the fact that taking average power readings each second is too coarse a frequency to capture the device's

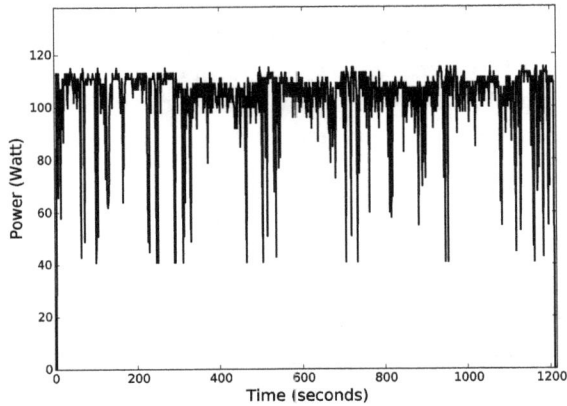

Fig. 7.4   An LCD TV exhibits a stable max behavior in its power usage.

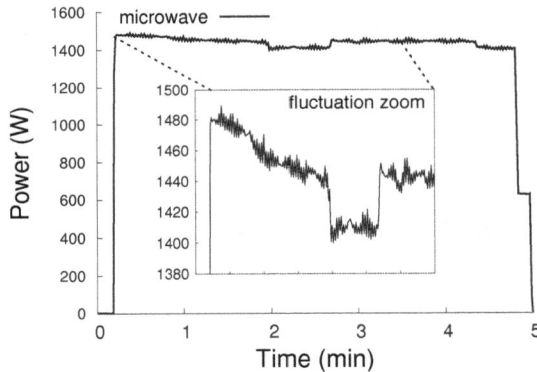

Fig. 7.5   A microwave exhibits random model with upper and lower bounds of power usage.

repetitive behavior. We model such loads by determining upper and lower power usage bounds, denoted by p_{max} and p_{min}. When active, our model randomly varies power within these bounds using a random walk. Note that the random range model is similar to the stable min-max model in that both employ upper and lower bounds on power usage. However, while the deviations in the stable min-max model are spikes from a stable value, those in the random range model are power variations within a range. The microwave is an example of a load that exhibits this behavior as shown in Figure 7.5.

Random range models require determining the minimum and maximum of the load's range of power usage. We determine these values by simply choosing the minimum and maximum power values observed in training data, or by deriving a distribution of power values from the data and choosing a high and low percentile of the distribution to be the minimum and maximum, p_{min} and p_{max}. We then model the variations with a random walk within the range.

7.3.3 *Compound Model Types*

While the models above accurately capture the behavior of simple loads, many loads, including large appliances, exhibit complex behavior from operating a variety of smaller constituent loads. We devise two types of compound models for complex loads that use the basic building blocks above.

7.3.3.1 *Cyclic Model*

Cyclic loads repeat one of the basic model types in a regular pattern, often driven by timers or sensors. A cyclic model augments a basic model by specifying the length of the active and inactive periods, t_{active} and $t_{inactive}$, each cycle. Constructing cyclic models is straightforward, since it only requires extracting the duration of the active and inactive periods from the empirical data. We currently use the mean of the active and inactive periods from the time-series observations to model t_{active} and $t_{inactive}$. Figure 7.6 shows the cyclic power usage pattern of a clothes dryer.

7.3.3.2 *Composite Model*

Composite loads exhibit characteristics of multiple basic model types either in sequence or parallel. Example composite loads include dryers, washing machines, and dishwashers, as shown in Figure 7.7. *Sequential composite loads* operate a set of basic load types in sequence; we model them as simple piecewise functions that encode the sequence of basic load models, including how long each load operates. For instance, a model for a dishwasher is a sequence of stages: modeled as the operation of the motor (wash stage), pump (drain stage), motor (rinse stage), pump (drain stage) and heater (dry stage), where each individual stage uses an inductive or resistive load. Some loads also exhibit characteristics of two or more basic models in *parallel* if

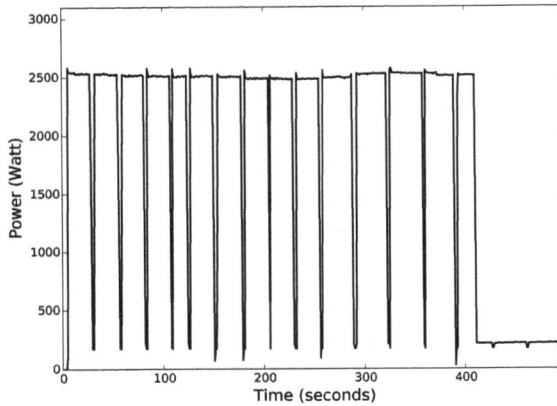

Fig. 7.6 A clothes dryer exhibits cyclic behavior in its power usage.

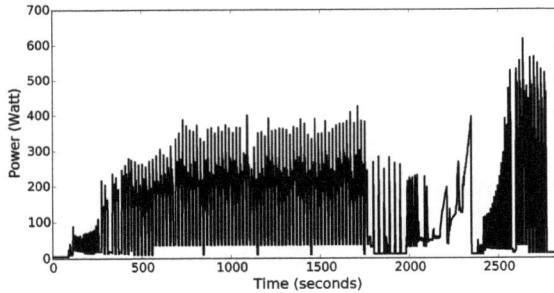

Fig. 7.7 A washing machine is a composite load that exhibits rich variations in its power usage. Each phase of the laundry wash cycle can be modeled using one of our basic load models.

two basic loads operate simultaneously. For example, a refrigerator may simultaneously activate both a compressor and an interior light. We model *parallel composite loads* by summing the power usage for two or more of the basic model types. Finally, composite loads may also be cyclic, referred to as *cyclic composite loads*, which repeat a pattern of individual model types at regular intervals.

Constructing load-specific composite models is more complex and requires additional manual inputs. For example, constructing a *sequential composite* model requires manually partitioning and isolating load time-series data into individual sequences that reflect the activation of the various load components. We must then construct basic models from above for each component in the sequence. The composite model is then simply a concatenation of these piecewise models in sequence.

As an example, Figure 7.7 shows an extended operating cycle of the washing machine with the annotations for different basic load model types in the sequence. We represent these models as large piecewise functions of the basic models describing each constituent load. Similarly, many of the large appliances that have composite models also have numerous operating states. For example, the washing machine and dryer in one of our homes have over 25 different types of cycles. Ideally, a model includes a different piecewise function for each cycle type. However, in the homes we monitored, we have found that most residents operate devices using only a few states — in most cases one.

Constructing parallel composite models poses many challenges. Since the time-series data for a load captures the power usage for all components that are concurrently active, there is no straightforward general-purpose technique to extract individual models from the composite time-series data. In practice, however, extracting basic models is often possible through exogenous means. For instance, many loads permit operating individual components to isolate them for profiling, e.g., such as running a dryer on tumble mode without any heat or using an air conditioner's fan without any cooling. After separately profiling a constituent load, such as the tumbler or fan, it is possible to operate the compressor and the fan, and then infer the compressor power usage by "filtering out" the tumbler or fan usage from the aggregate. In some devices, such as a refrigerator, it also might be possible to deploy additional sensors that monitor important events, such as a door opening that triggers lights, to filter them out. Ideally, the model of a complex composite device would be provided by the device manufacturer, as the problem of identifying the *components* of a composite device is largely orthogonal to the problem of *modeling* each component. However, using the techniques described previously, it is generally possible to identify the key components even without detailed knowledge of the internals of the device.

It is noteworthy that the above modeling framework can also be used by higher-level inference and analytic approaches to implement various energy management and optimization tasks.

7.4   Summary and Takeaways

The two topics discussed in this chapter are apparently different, but they are connected in the sense that both the topics make a case for consumers' participation in energy management. While brownout is a more welcoming option for the consumer than rolling blackouts, NILM techniques can help utilities in influencing consumers to be more energy conscious. From the perspective of power utilities — these technological solutions can improve consumers' satisfaction level and make them grid-friendly.

Appendix A

Electrical Energy: Some Basic Concepts

We now give a a bird's eye view of some concepts needed to appreciate why we need a smarter management of electrical energy.

In most cases electrical energy is produced as a result of conversion from mechanical energy. The process involves a magnetic field and its interaction with *conductor(s)* of electric current. Electricity generators convert mechanical energy into electrical energy where the mechanical energy required to rotate the prime mover comes from i) burning of fossil fuels (Thermal Power Plants — TPP, gas/diesel turbines), ii) utilization of potential energy of water stored at a height (Hydro-electric/Hydel power plants) or iii) controlled fission in a nuclear reactor (Nuclear Power Plants — NPP). In contrast, solar power is generated directly from sunlight using photo-voltaic cells.

When there is a relative motion between a magnetic field B and a conductor and the conductor cuts through the magnetic flux lines, an Electro-Motive Force (*emf*) E is generated due to electromagnetic induction causing a potential difference (voltage v) between the terminal points of the conductor.

If a conductor moves (relative motion) with a velocity V, the emf induced E in an element dl of a conductor is expressed as

$$E = (\vec{V} \times \vec{B}).\vec{dl}$$

The generated voltage, which is denoted as v (same as E), will produce current i if the circuit is closed through a load. Note that this generation can be DC or AC depending on the variations in B that the conductor observes due to the relative motion.

A generator that feeds an electric grid, a network of multiple generators connected in parallel, is a synchronous generator, where a DC magnetic field is rotated with the help of a *rotor* at a fixed angular speed ω_s, called synchronous speed. The prime mover that rotates the rotor can be a steam turbine or a hydro-turbine. Emf is induced in the stationary conductors placed in the *stator* due to the relative motion between the conductor and the magnetic field. Specific spatial distribution of conductors on the cylindrical stator and suitable connections among the group of three conductors produce three phase ($3Ph$) sinusoidal AC voltages. Note that single phase ($1Ph$) sinusoidal AC voltage can be generated if a conductor coil or a group of conductors connected in series forming a coil is rotated at a constant speed in a constant magnetic field.

A.1   Power Consumption and Loads in AC Circuit

Other than the resistive heating or incandescent lighting, most of the loads, industrial or domestic, are mechanical, e.g., a big industrial motor driving a pump, or a small motor driving the mixer-grinder in our homes. Here, electrical energy is supplied to these devices or equipment and it gets converted into mechanical energy driving the loads. The conversion of electrical to mechanical energy involves the generation of mechanical force F when a current carrying conductor is placed under the magnetic field B and is expressed as

$$\vec{F} = i\vec{dl} \times \vec{B},$$

where, i is the current through the small length element dl of the conductor.

Three phase ($3Ph$) or single phase ($1Ph$) loads, when connected to corresponding supplies, draw sinusoidal current and there is a phase difference between the supply voltage v and current i, if the load is not purely resistive. Loads are usually inductive and the current i lags behind the voltage v.

Let us now discuss the basic types of loads and how they draw current when connected to an AC voltage source. The loads are basically three types viz., *resistive, inductive* and *capacitative*. However, real-world loads, though mostly resistive and inductive, can be any combination of these three types of basic loads. Here, we will introduce the voltage-current relationship when sinusoidal voltage is connected to these basic load types. In order to do so, first we will explain the terms AC voltage v, current i and frequency f and how they are represented in a phasor diagram.

A sinusoidal voltage source v is described as a function

$$v = V\cos \omega t$$

where, lowercase v is the instantaneous voltage, uppercase V is the maximum voltage and ω is the angular velocity of rotation of the coil (or rotor of synchronous generator). This ω determines the frequency of the AC source, known as supply frequency.

$$\omega = 2\pi f$$

In India and most of the countries, the commercial supply frequency is 50 Hz, while in USA and Canada $f = 60$ Hz.

Similarly, AC current i is represented as

$$i = I cos\ \omega t$$

where, i represents instantaneous current, I the maximum current.

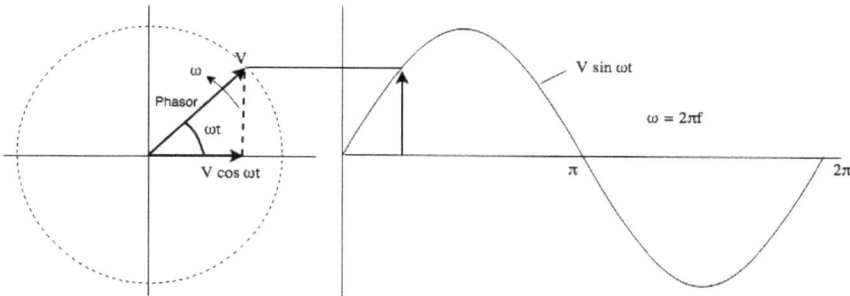

Fig. A.1   Phasor diagram: rotating vector representing sinusoidal voltage.

In order to represent sinusoidal voltage and current rotating vector the diagram in Figure A.1 is used. Observe that in Figure A.1, the instantaneous value (at any time t) of the parameter (voltage), which varies sinusoidally is represented by the projection of the rotating vector (V) on the vertical axis. The length of the projection is equal to the amplitude of the parameter at time t.

A rotating vector is called *Phasor* and we will consider its *projection on the horizontal axis* to represent the instantaneous value of the parameter for the *convenience of representation* in a phasor diagram. Note that a sine wave becomes a cosine wave when shifted by a phase angle of $\pi/2$. A phasor diagram is a convenient way to represent sinusoidal voltage, current, etc. and their phase differences. Further, a vectorial representation of sinusoidal voltage and current with phase differences facilitates the addition of these quantities using simple vector addition.

(a) Resistive circuit

(b) Current and voltage waveform

(c) Phasor diagram

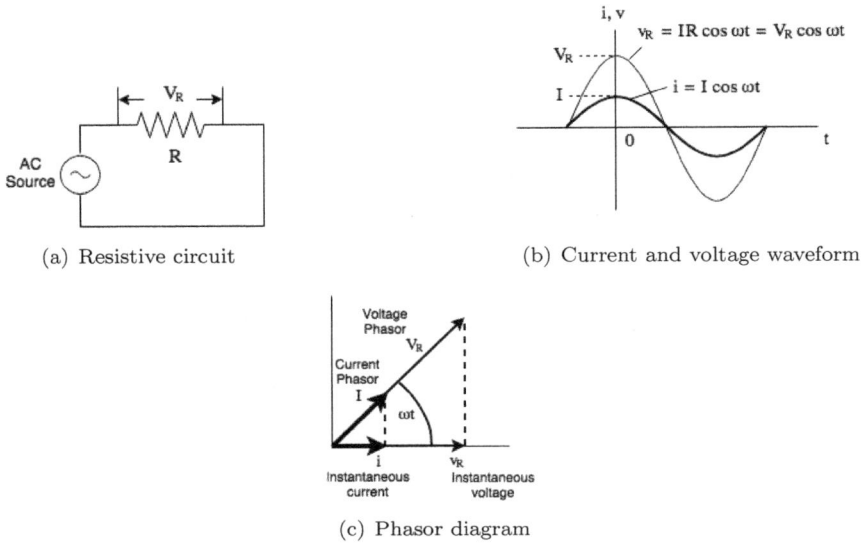

Fig. A.2   Resistive load connected to AC voltage.

Resistive load in AC circuit:

Consider Figure A.2(a) showing a resistance connected to an AC voltage. The instantaneous current flowing through the resistance R and the voltage v_R across R can be obtained from Ohm's law.

$$v_R = iR = IR \ cos \ \omega t$$

where, I represents the maximum value or the *amplitude* of current i and $i = I \ cos \ \omega t$. The *amplitude* of voltage $V_R = IR$ and therefore, we can also write

$$v_R = V_R \ cos \ \omega t$$

We observe from the voltage and current waveforms shown in Figure A.2(b) that both v and i are in phase and have same frequency. In phasor terms, both the voltage and the current phasors are in parallel at any instant of time and rotate together at same frequency as shown in Figure A.2(c). The instantaneous values of v and i, as shown in Figure A.2(c), are the projections of the respective phasors on the horizontal axis.

(a) Inductive circuit

(b) Current and voltage waveform

(c) Phasor diagram

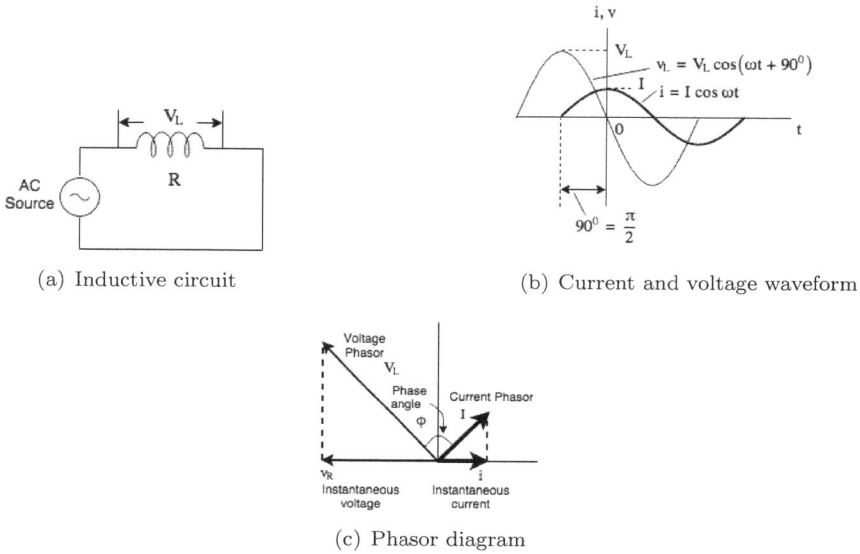

Fig. A.3   Inductive load connected to AC voltage.

Inductive load in AC circuit:

Let us now connect a pure inductor L to the same AC voltage source as shown in Figure A.3.

The current that flows through the inductor L is $i = I \cos \omega t$. The elementary principle of electrical circuit theory tells us that when a time-varying current i flows through an inductor L, a *self-induced emf* ϵ is produced and a potential difference V_L across the inductor is observed such that

$$\epsilon = -L \frac{di}{dt}$$

and $v_L = -\epsilon$, as the induced emf is in the direction of the current. So,

$$v_L = L \frac{d}{dt}(I \cos \omega t) = -I\omega L \sin \omega t$$

We can rewrite the above equation as

$$v_L = I\omega L \cos (\omega t + 90°)$$

$$v_L = V_L \cos (\omega t + 90°) \tag{A.1}$$

where $V_L(= I\omega L)$ denotes the amplitude of the inductor voltage.

(a) Capacitative circuit

(b) Current and voltage waveform

(c) Phasor diagram

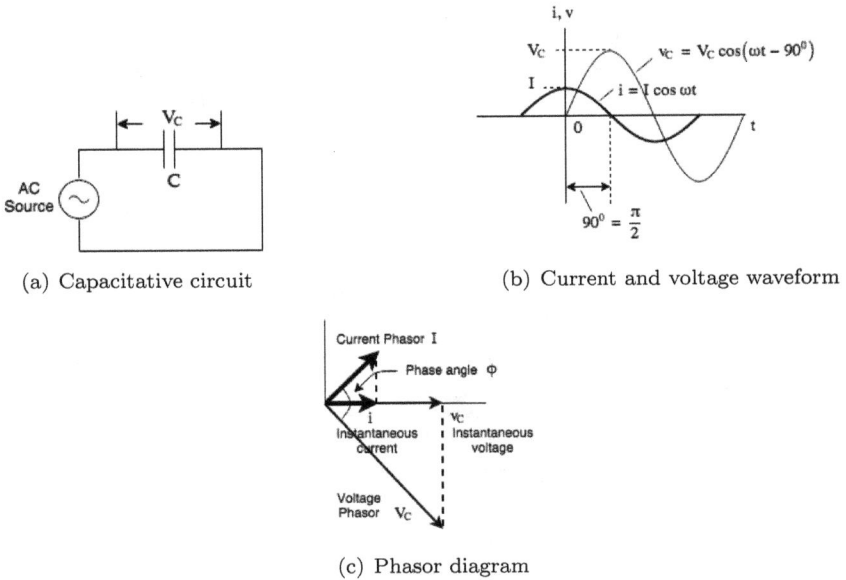

Fig. A.4   Capacitative load connected to AC voltage.

V_L is also written as

$$V_L = IX_L \tag{A.2}$$

where $X_L = \omega L$ is called the *inductive reactance*. Also, note that current amplitude $I = V_L/X_L$.

Equation (A.1) shows that the inductive voltage v_L is 90° out of phase with the current $i = I \cos \omega t$, and the voltage *leads* the current, as shown in Figure A.3(b). Observe in Figure A.3(b) that the peak of voltage waveform is 90° ahead (*lead*) of the peak of the current waveform when compared to no phase difference between v and i in the case of pure resistive load as in Figure A.2(b). The corresponding phasor diagram is shown in Figure A.3(c).

Capacitative load in AC circuit:
Let us now replace the inductor with a capacitor C as shown in Figure A.4(a).

In this circuit, current $i = I \cos \omega t$ flows through the capacitor. When an AC voltage is applied across a capacitor, the capacitor gets charged and discharged in every cycle due to the alternating nature of the voltage applied across it and at each instant a flow of *displacement current* takes place through the capacitor plates as if charge is being conducted through the capacitor.

In order to find the instantaneous voltage v_C across the capacitor, let us recall from elementary physics that current is defined as

$$i = \frac{dq}{dt}$$

where charge is denoted by q. Again $i = I\ cos\ \omega t$. So,

$$\frac{dq}{dt} = I\ cos\ \omega t$$

Integrating,

$$q = \frac{I}{\omega}\ sin\ \omega t \tag{A.3}$$

Recall again the elementary physics lessons on electricity, the voltage v_C across a capacitor with capacitance C is related to charge q as

$$C = \frac{q}{v_C}$$

Using this relationship in Equation (A.3), we get

$$v_C = \frac{I}{\omega C}\ sin\ \omega t$$

Rewriting,

$$v_C = \frac{I}{\omega C}\ cos(\omega t - 90°) \tag{A.4}$$

We define $X_C = 1/\omega C$, as the *capacitative reactance* so that

$$v_C = IX_C\ cos(\omega t - 90°) = V_C\ cos(\omega t - 90°)$$

Therefore, the amplitude of capacitative voltage is $V_C = IX_C$.

Equation (A.4) shows that the capacitative voltage v_C is 90° out of phase with the current $i = I\ cos\ \omega t$, and the voltage lags the current, as shown in Figure A.4(b). The corresponding phasor diagram is shown in Figure A.4(c).

General Voltage and Current Phasor Relationship:

As discussed earlier, real-world loads are not always purely restive or inductive and mostly loads are combinations of resistance and inductance. Therefore, the phase angle between voltage and current varies depending on the values of resistance and inductive reactance.

If we consider a general case where $i = I\ cos\ \omega t$ and $v = V\ cos\ (\omega t + \phi)$ at any time instant t, then the *phase angle* between voltage and the current is ϕ. The value of ϕ can vary from $-90°$ to $+90°$.

Active (Real), Reactive Power and Apparent Power:

In case of DC, the power P supplied to the load is always the active or real power and simply expressed as

$$P = vi$$

In case of AC circuit, due to the phase difference[1] of angle ϕ between the sinusoidal voltage and current, the *active power* W — voltage v times the real component of current i (in phase with v), which is the mechanical output available to drive any load is

$$W = vi\cos\phi \qquad (A.5)$$

The *reactive power* or watt-less power Q is v times the orthogonal component (with respect to v) which does not produce any useful work. This is the power that goes back and forth between the generator and the load, so that the reactive power demand by the load is always met by the generator.

$$Q = vi\sin\phi$$

The *apparent power* S is the combined active and reactive power

$$S = vi \quad \text{in } VA \text{ (volt-ampere) units}$$

This term is used to express the rating of a generator, usually in KVA (Kilo Volt-ampere) or MVA (Mega Volt-ampere).

Power Factor — From Equation (A.5), it is clear that with lower phase angle (ϕ) between voltage and current, we get higher active power. The entity $Cos\ \phi$ is known as the *power factor (pf)* and a higher *pf* is always desirable.

---

[1] In case of purely resistive load, the current i is in phase with voltage v i.e., $\phi = 0$.

Appendix B

Short Introduction to Power System Stability

Power system stability may be defined [William D. Stevenson (1982)] as the property of the system which enables the synchronous machines (forming the grid) of the system to respond to a disturbance from normal operating conditions so as to return to a condition where their operation is again normal. A generator is said to be *synchronised* with the grid when the voltage, frequency and phase sequence of both the generator and the grid are the same. Thus, we can also define the power system stability as the ability of the power system to return to steady state, following a disturbance due to variation in generation, and load including the occurrence of faults, without losing synchronism.

The stability of power system is categorised as *steady state*, *transient* and *dynamic stability*. The steady state and dynamic stability studies involve one or just a few machines undergoing slow or gradual disturbances in the operating conditions. The *transient stability* is the system's ability to remain in synchronism following major disturbances like sudden load changes, loss of generating unit(s), transmission system faults or line switching. Transient stability analysis enjoys greater importance due to its wider applicability in practice.

When there is any such disturbance, the speed of the synchronous generator's rotor departs from its synchronous speed, momentarily. The stability analysis tells us if the rotors of the machines being perturbed would return back to its constant speed (synchronous speed) operation.

Successful operation of the synchronous machine (generator or motor) demands equality of the mechanical speed of the rotor and the speed of stator magnetic field [Fitzgerald *et al.* (1985)]. Synchronising forces tend to maintain this equality whenever this relationship is disturbed. Under steady state, the rotating masses, i.e., the rotor with its prime mover[2] are in synchronism with all other machines operating at synchronous speed in the power system. The maximum power a synchronous machine can deliver depends on the maximum torque which can be applied without loss of synchronism. Thus, stability depends on the short-time overload (or load withdrawal) that a machine can withstand by adjusting[3] (increasing or decreasing) the torque that can be delivered without loss of synchronism. In this context, the reader may have a relook at the simple example in Section 2.4.2 of Chapter 2 explaining the basic concept behind the balance in generation and consumption maintained in an electric network.

B.1 Rotor Dynamics and Swing Equation

The rotor of a synchronous machine rotates at an angular speed ω_m, which is also its synchronous speed. When a synchronous generator is connected to the grid, current flows in the stator winding (to drive electrical load connected to the grid) and a rotating magnetic field is produced in the stator.[4] There is some relative angular displacement between the rotor axis and the axis of rotation of the stator magnetic field, known as the *load angle* or *torque angle* δ which is directly proportional to the loading (electrical power demand) on the generator.

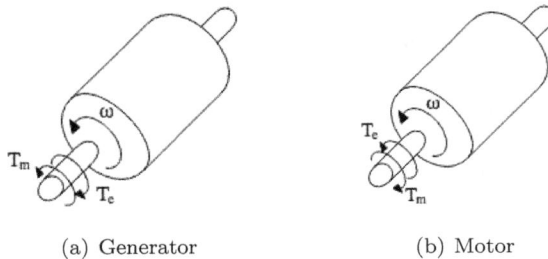

(a) Generator   (b) Motor

Fig. B.1   Torques acting on the rotor of a synchronous machine and their directions of rotations in a generator and a motor.

---

[2]The prime mover can be steam or hydro turbine.
[3]For example, in the case of a steam generator, the steam supply to the prime mover is controlled.
[4]Due to the specific spatial distribution of the 3-phase stator winding and the sinusoidal variation in the rotor magnetic field as seen by the stator winding (conductor) because of the rotation of the rotor.

Two kinds of torque work on the rotor of a synchronous machine, the mechanical torque T_m and the electrical torque T_e. In case of a generator, see Figure B.1(a).

- The mechanical torque T_m supplied on the shaft of the rotor by the prime mover, which acts to accelerate the rotor and
- The electromagnetic or electrical torque T_e, which acts to decelerate the rotor, and accounts for the total output power of the generator.

In case of a motor (see Figure B.1(b)), we get

- The electrical torque T_e that drives the rotor to rotate and
- The mechanical torque T_m, due to the load connected to the rotor, acts against the electrical torque to decelerate the rotor.

The electrical torque T_e is the result of the force experienced by the rotor due to the interaction between magnetic fields of the rotor and the stator. We will discuss the torque balance in a generator as it is the generators, which are controlled by the utilities to meet the power requirement of the loads at the consumer end. Note that considering the rotor speed and its magnetic field excitation constant, the magnitude of the stator magnetic field depends on the current flowing in the stator conductors, which in turn depends on the load that the generator is feeding. Thus, T_e is related to the output power of the generator.

Under steady-state operation, the generators T_m and T_e are equal and the rotor rotates at its designed synchronous speed. In case of temporary imbalance between these two torques due to disturbances, the rotor will either accelerate or decelerate momentarily. This acceleration torque represented by T_a can be expressed as

$$T_a = T_m - T_e \tag{B.1}$$

The directions of T_m and T_e are opposite as shown in Figure B.1.

Let θ_m denote the absolute measurement of rotor angle with respect to a stationary reference axis on stator. Therefore, θ_m will increase continuously with time even at constant synchronous speed. Since the rotor speed relative to the synchronous speed is of interest, it is convenient to measure the rotor angular position with respect to a reference axis that rotates at synchronous speed ω_s.

So, we define

$$\theta_m = \omega_s t + \delta \tag{B.2}$$

Note that the reference axis here is the axis of rotation of the stator magnetic field and δ is the *load angle* or *torque angle* as discussed earlier in this section.

Taking derivative,

$$\frac{d\theta_m}{dt} = \omega_s + \frac{d\delta}{dt} \tag{B.3}$$

and

$$\frac{d^2\theta_m}{dt^2} = \frac{d^2\delta}{dt^2} \tag{B.4}$$

Equation (B.3) shows that only when $d\delta/dt$ is zero, the rotor angular velocity is constant at ω_s and $d\delta/dt$ (measured in mechanical radian per second) represents the deviation of rotor speed from synchronism.

The accelerating torque T_a of the rotor can be expressed, using the elementary principle of dynamics as

$$J\frac{d^2\theta_m}{dt^2} = T_a$$

where J denotes the total moment of inertia of the rotor mass in $Kg\text{-}m^2$.

Substituting T_a in Equation (B.1) and using Equation (B.4)

$$J\frac{d^2\delta}{dt^2} = T_m - T_e \quad N\text{-}m \tag{B.5}$$

Now, angular rotor velocity ω_m can be expressed as

$$\omega_m = \frac{d\theta_m}{dt}$$

From the elementary principles of dynamics, we know that power equals torque times the angular velocity. Therefore, the accelarating power P_a can be expressed as

$$P_a = J\frac{d^2\delta}{dt^2} \times \omega_m = P_m - P_e$$

where P_m denotes the shaft power input by the prime mover and P_e the electrical power output.

$$J\omega_m\frac{d^2\delta}{dt^2} = P_m - P_e \quad W \tag{B.6}$$

Equation (B.6) is called the *swing equation*.

We can consider the rotor speed and therefore the shaft power P_m as constant because the rotor speed does not change till the time the governor acts to change the turbine (prime mover) speed. Therefore, the acceleration or deceleration of the rotor will depend on the electrical power output P_e during the transient phase.

Let us consider a case where initially a generator is in synchronism with the other machines in the power system and an electrical load is disconnected. This will reduce the electrical power output P_e causing the rotor to accelerate and δ will increase till the steam governor takes corrective action to reduce the mechanical power input P_m by reducing the steam supply. But, even after the reduction of steam supply, the inertia of the rotor mass will not allow the rotor to come back immediately to its synchronous speed. This will cause oscillations before the machine comes back to a steady state[5] for a new operating condition balancing the new electrical power demand P_e with the required P_m. If an additional load is connected, the P_e will increase causing the rotor to decelerate momentarily and the governor will take corrective action to match the increased demand of P_e from the generator.

In case of fault, the protection system acts to isolate the generator from feeding the fault current by opening the right circuit breaker in a short period of time. So, the electrical power output P_e from the generator comes to zero. This causes acceleration in the rotor and thus δ increases. When the fault is cleared, the P_e demand increases abruptly, but the increase in δ does not stop immediately due to rotor inertia. This causes oscillation in the rotor but eventually comes to a steady state as the system is designed to be damped.

The stability analysis tells us how much variations in δ is acceptable before corrective action takes place.

B.2   Power Flow and Power Angle Equation

Changes in the generator power output P_e depends on the conditions on the network of the generators and the connected loads on the power system. Disturbances resulting from large load changes, network faults, circuit breaker operation can cause rapid or sudden change in P_e creating electromechanical transient.

---

[5]Note that *damper windings* are provided in the rotor to damp out oscillations of the rotor about its equilibrium position.

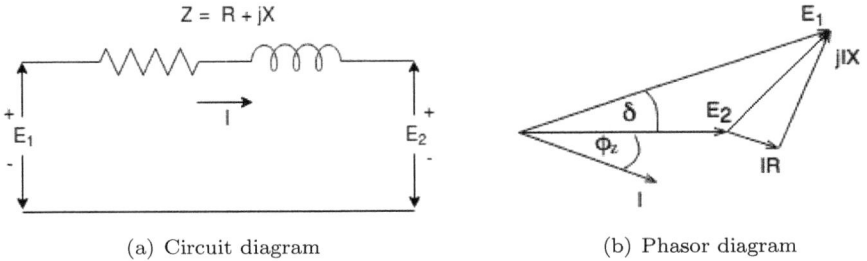

(a) Circuit diagram   (b) Phasor diagram

Fig. B.2   Power angle: two voltages connected by impedance.

Let us consider the simple circuit of Figure B.2 showing two AC voltages connected by an impedance. The power flow through this impedance can represent both the cases: i) power output from a generator as the machine can be represented by an impedance (synchronous impedance) connected to the induced stator voltage E_1 at one end and the terminal voltage E_2 at the other end (the infinite bus voltage), and ii) power flow in a transmission line with two generators connected one at each end, where the transmission line is represented by an impedance that can include impedances of the line and the transformer bank.

Figure B.2(a) shows two AC voltages E_1 and E_2 connected by an impedance $Z(= R + iX)$. The current flowing through Z is I. The power delivered to E_2 end is

$$P_2 = E_2 I \cos \phi_z \qquad (B.7)$$

where ϕ_2 is the phase angle by which the current I lags behind E_2 as shown in the phasor diagram,[6] Figure B.2(b) of the circuit under discussion. The phasor current is

$$I = \frac{E_1 - E_2}{Z}$$

In polar form

$$I = \frac{E_1 \angle \delta - E_2 \angle 0}{Z \angle \phi_z}$$

---

[6]Note that voltage drop IR across the resistance R is in phase with current I and the drop across the inductance is jIX, which is 90° out of phase (represented by the jth term) with the current I, as shown in Figure B.2(b). Also, $\vec{E_1} = \vec{E_2} + \vec{I}\vec{Z} = \vec{E_2} + \vec{I}\vec{R} + j\vec{I}\vec{X}$.

$$I = \frac{E_1 \angle (\delta - \phi_z)}{Z} - \frac{E_2 \angle (-\phi_z)}{Z}$$

where δ is the phase angle by which voltage E_1 leads the voltage E_2.

The real component of the phasor equation is the component of I in phase with E_2 expressed as

$$I \cos \phi_2 = \frac{E_1}{Z} \cos(\delta - \phi_z) - \frac{E_2}{Z} \cos(-\phi_z)$$

Substituting $I \cos \phi$ in Equation (B.7), we get

$$P_2 = \frac{E_1 E_2}{Z} \cos(\delta - \phi_z) - \frac{E_2^2}{Z} \cos(-\phi_z)$$

Now as $z = R + jX$, $\cos(-\phi_z) = \cos(\phi_z) = \frac{R}{Z}$.
Therefore,

$$P_2 = \frac{E_1 E_2}{Z} \cos(\delta - \phi_z) - \frac{E_2^2 R}{Z^2} \tag{B.8}$$

If we take $\alpha_z = 90° - \phi_z$, then

$$P_2 = \frac{E_1 E_2}{Z} \sin(\delta + \alpha_z) - \frac{E_2^2 R}{Z^2} \tag{B.9}$$

Also note that $\alpha_z = tan^{-1} \frac{R}{X}$, whose value is very small, due to very small value of the resistance R as compared to the reactance X. For practical cases, R is negligible. It follows that

$$P_2 = \frac{E_1 E_2}{X} \sin \delta \tag{B.10}$$

Equation (B.10) shows that if the resistance is negligible, the maximum power flow occurs when $\delta = 90°$ and the value is

$$P_{max} = \frac{E_1 E_2}{X}$$

It may be noted that this power angle δ corresponds to the torque angle discussed in Equation (B.2). However, the comparison is beyond the scope of this book.

Power Flow Direction:
It can also be observed from Equation (B.10) that power flows from E_1 end to E_2 end of the transmission line (or from generator to the bus) if E_1 leads E_2 by some phase angle δ. If otherwise, power will flow from E_2 end to E_1 end.

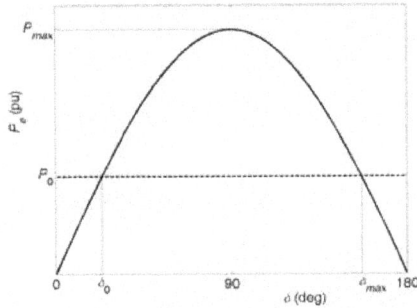

Fig. B.3   Power angle curve.

Power Angle Curve:

Figure B.3 shows the plot of power angle Equation (B.10). The mechanical power input to the generator P_m is constant and intersects the sinusoidal power angle curve at two points as can be observed from Figure B.3. So, for a given power P_0, there are two possible values of the power angle — δ_0 and δ_{max}, where $\delta_{max} = 180° - \delta_0$.

Any disturbances like sudden change in load or fault in the transmission line will cause a change in the power angle δ due to imbalance between P_m and P_e. Therefore, corrective action is required to be taken within a short period of time before synchronism is lost. In case of fault, the fault has to be identified and cleared within a short period termed as *fault clearing time* so that the deviation in δ is within limits allowing the system to go back to its equilibrium state.

Appendix C

Thermodynamic Principles and RC-Modeling of Buildings

When we talk of thermal comfort and thermal conditioning resources, like air-conditioners, heaters, dehumidifiers, we deal with the principles of thermodynamics, electrical engineering and often refer to the related Industry Standards.

In this appendix, we present i) some basic concepts of thermodynamics and ii) the popular equivalent RC modeling of a building space, which we believe will be helpful to the readers in two ways — i) Developing/Refreshing some basic concepts for a better reading and ii) exploring the implementation aspects of the ideas presented in this monograph.

C.1   Basic Concepts of Thermodynamics

The basic concepts of heat transfer and thermodynamics essential to understand the working of air-conditioners, dehumidifiers and thermal conditioning of a closed space are presented here.

Fundamental Principles of Thermodynamics behind AC

The following fundamental principles of thermodynamics are presented here which will help us in understanding the working of an air-conditioner (presented above) better.

First Law of Thermodynamics (Conservation of Energy): The process of heat transfer is the transfer of energy between two bodies. If we add Q amount of heat energy into a system, the system may i) do some work as in the case of an automobile engine, where heat energy produced by burning fuel gets converted into work, or ii) may not do any work as in the case of heating water. When the system does no work after Q amount of heat is added, its internal energy U increases and the increase in internal energy ΔU is equal to Q, i.e., $\Delta U = Q$. If the system does some work W without addition of any heat, then its internal energy decreases by ΔU. This is observed when the hot and compressed refrigerant expands through a throttle valve in an AC. Alternatively, if work is done on the system by adding energy from an external source, its internal energy is increased. This happens, when the compressor compresses the refrigerant, its temperature increases due to increase in the internal energy. It follows that

$$\Delta U = -W \qquad (C.1)$$

Therefore, considering the general case, when both heat transfer and work take place, the change in internal energy ΔU is expressed as

$$\Delta U = Q - W$$

where, Q = heat added/removed and W = work done on/by the system. This is the first law of thermodynamics, which is based on the principle of conservation of energy. In case of infinitesimal change in the state

$$dU = dQ - dW \qquad (C.2)$$

Cooling on Adiabatic Expansion (ideal gas):
It is the process of adiabatic expansion of refrigerant through throttle valve, which brings down its temperature before it goes to the evaporator coil. The air (water in case of chiller plant) when comes in contact with evaporator coil, it releases heat into the refrigerant and therefore gets cooled.

In an adiabatic process of expansion or compression, no heat flows into or out of the system. This phenomenon usually occurs when a gas is expanded or compressed so rapidly that there is no time for heat to flow in or out.

For an ideal gas, the change in internal energy under any thermodynamic process is obtained [Young and Freedman (2008)] by

$$dU = nC_v dT \qquad (C.3)$$

where, n is the number of moles of the gas, C_v is the molar heat capacity of the gas at constant volume and dT denotes the change in temperature.

Further, work done in the process where the volume changes without any change in pressure p, is expressed as

$$dW = PdV \qquad (C.4)$$

From the First Law,

$$dW = -dU \qquad \text{as } dQ = 0 \text{ for adiabatic process} \qquad (C.5)$$

Therefore,

$$nC_v dT = -PdV \qquad (C.6)$$

From the ideal gas equation $PV = nRT$ (R is the ideal gas constant)

$$nC_v dT = \frac{nRT}{V} dV \qquad (C.7)$$

Considering the fact that $C_p = C_v + R$, we can derive the following from Equation (C.7), by simple integration

$$T_1 V_1^{\gamma-1} = T_2 V_2^{\gamma-1} \qquad (C.8)$$

where, $\gamma = C_p/C_v$ denotes the ratio of heat capacities, T_1 and T_2 denote the initial and the final temperature, respectively and the corresponding volumes are represented by V_1 and V_2.

Therefore,

$$T_1 = (V_2/V_1)^{\gamma-1} T_2 \qquad (C.9)$$

If adiabatic expansion takes place i.e., if $V2/V1 > 1$, then we get

$$T_2 < T_1 \qquad (C.10)$$

It shows that cooling occurs in the adiabatic expansion process.

C.2 Principles of Thermodynamics and RC Modeling of Building Space

The analogy between a thermal system and an electrical network and how is analogy used in the development, can be the RC modeling of building space as described in this section.

C.2.1 *Thermal Parameters and its Electrical Analogs*

C.2.1.1 *Conduction*

When a quantity of heat dH is transferred through a material in a time dt, the rate of heat flow is dH/dt. We call this as *heat flux* Q. From the elementary principles of heat transfer physics and from the experimental results it is known that heat flux is proportional to the cross-sectional area A and to the temperature difference $(T_2 - T_1)$ between the two ends of the material and inversely proportional to the length x.

$$Q = \frac{dH}{dt} = kA \frac{T_2 - T_1}{x} \qquad (C.11)$$

where k is the proportionality constant called the thermal conductivity of the material.

This can be rewritten as

$$Q = \frac{1}{R}(T_2 - T_1) \tag{C.12}$$

where, R is the thermal resistance of the material.

Comparing this with Equation (C.11), we can see that

$$R = \frac{x}{kA} \tag{C.13}$$

This is analogous to the electrical resistance R offered to the flow of charge by a material of length x having a cross-sectional area A, which is mathematically represented as

$$R = \rho\frac{x}{A} = \frac{x}{\sigma A} \tag{C.14}$$

where, ρ is the resistivity of the material and the electrical conductivity is represented by $\sigma = 1/\rho$.

Ohms law states that the voltage v across a resistor is directly proportional to the current i flowing through it.

$$v = iR \tag{C.15}$$

where R is the proportionality constant known as the resistance, a material property as stated in Equation (C.14) and the voltage v is the potential difference between the two end points of the material. Therefore,

$$v_2 - v_1 = iR \tag{C.16}$$

where, v_1 and v_2 are the voltages at the two end points of the conductor (resistor). Further, it is known from the elementary physics of electricity that current i is the rate of flow of charge q. So, using Equation (C.16), we can express current i as

$$i = \frac{dq}{dt} = \frac{1}{R}(v_2 - v_1) \tag{C.17}$$

This is analogous to the thermal conduction Equation (C.12). Equations (C.17) and (C.14) together offer electrical analogs to the heat flow and thermal resistance as tabulated in Table C.1.

C.2.1.2 *Energy Storage*

When heat energy is transferred to/from a material, its temperature T changes. It is the storage/loss rate of energy Q that effects the change in the temperature and can be defined by

$$Q = C\frac{dT}{dt} \tag{C.18}$$

where, C is thermal capacitance of the material.

Similarly, the amount of charge q stored in a capacitor is directly proportional to the voltage applied to it, so that

$$q = Cv \tag{C.19}$$

where C is the proportionality constant known as the *capacitance*. Differentiating,

$$\frac{dq}{dt} = C\frac{dv}{dt}$$

$$i = C\frac{dv}{dt} \tag{C.20}$$

Comparison of this Equation (C.20) with Equation (C.18) offers an electrical analog of the thermal storage as presented in Table C.1.

C.2.1.3 *Convection*

The convective heat transfers that can take place in a building are

(1) Between a wall surface at temperature T_w and the room air volume at T_i.
(2) Between the evaporator coil surface of the in-door unit (IDU) of an air-conditioner at temperature T^E and the room air volume at T_i in case of cooling.
(3) Between the heating coil surface of a space heater at temperature T^E and the room air volume at T_i in case of heating.

The convective heat flow exchange can be derived using Newton's law [Nelkon and Parker (1977); Fraisse *et al.* (2002)] by

$$Q = h_c S(T_s - T_i) \tag{C.21}$$

where, h_c is the convective heat transfer coefficient, S is the surface area and T_s is the generic representation of the temperature of the heating/cooling surfaces stated above. Thus the convective resistance is defined by

$$R = \frac{1}{h_c S} \tag{C.22}$$

C.2.1.4   *Radiation*

Exchange of long wave (LW) radiative heat occurs between the surface of a wall/window and the environment or the ambient.

Additional heat is absorbed in a room when a window/wall is directly exposed to the sun. For a window, this solar heat gain Q^s can be expressed [Skruch (2015)] as

$$Q^s = f^s f^g S^w H^s \qquad (C.23)$$

where, f^s is the shading factor, f^g is the glass solar factor and H^s is the incident heat flow per square meter of wall/window area S^w.

The radiative heat exchange [Young and Freedman (2008)] between two bodies is expressed as

$$Q = Se\sigma(T_2^4 - T_1^4)$$

where S is the surface area, e is the emissivity, T_2 and T_1 are the temperatures of the surfaces involved in radiative heat exchanges and σ is the Stephen-Boltzmann constant.

Based on the linearization of the fourth power law of radiation discussed in [Davies (1984)] the radiative component[7] of heat flux due to the Sun can be expressed as

$$Q = Se\sigma T_{av}^3(T_2 - T_1) = h_r(T_2 - T_1) \qquad (C.24)$$

where T_av is the average of T_2 and T_1, and h_r represents the overall radiative heat transfer coefficient.

In case of heat flux through a window/wall due to LW radiation, the notion of *sky temperature* T_{sky} is discussed in [Fraisse *et al.* (2002)] offering a linearized expression for the radiative component of heat transfer through the window/wall. This is done such that an equivalent RC network model can be developed to capture the radiative heat flux through the walls/windows. In [Fraisse *et al.* (2002)], external environment is also considered as a radiative heat source/sink. Thus the radiative heat flux Q^s through the window/wall is computed as

$$Q^s = h_{r_{vc}}(T_{sky} - T_w) + h_{r_{env}}(T_{env} - T_w) \qquad (C.25)$$

where, T_w is the window/wall temperature, the sky is represented by a hemisphere and the environment by a horizontal surface with $h_{r_{vc}}$ and $h_{r_{env}}$ as their respective radiative heat transfer coefficients. Considering shading factor and glass-solar factor for direct sunlight on wall/window,

$$Q^s = f^s f^g h_{r_{vc}}(T_{sky} - T_w) + h_{r_{env}}(T_{env} - T_w) \qquad (C.26)$$

---

[7]The other one is the convective component.

C.2.1.5   *Ventilation*

Exchanges of heat energy take place in a building due to ventilation. Based on the assumptions of constant ventilation and infiltration air flow-rate, the heat loss/gain due to ventilation can be expressed as

$$Q = c\rho v(T_a - T_z) \tag{C.27}$$

where, c is the specific heat capacity of air, ρ is the air density, T_a is the ambient temperature, T_z is the inside temperature of a room/zone.

In case of heating, if there exist a heat recovery system, the heat loss can be modified as

$$Q = (1 - \eta)c\rho v(T_a - T_z) \tag{C.28}$$

where η is the heat recovery efficiency.

C.2.2   *Equivalent RC Modeling*

The technique of resistance-capacitance (RC) network as an equivalent thermal model of a building is a widely used one. This technique is based on the heat transfer physics and its mathematical analogy in electrical RC network is as discussed above and summarized in Table C.1

It may be noted that in addition to the assumptions discussed in Section 4.4.1, the following assumptions are important considerations in RC modeling.

(1) The convection heat transfer coefficients remain constant and the effect of wind speed, surface-air temperature differences are negligible.

Table C.1   Thermal system and its electrical analogs.

Thermal System Parameters			Electrical System Parameters		
Quantity	Symbol	Unit	Quantity	Symbol	Unit
Heat Flux	Q	calorie/sec.	Current	i	Ampere
Temperature	T	degree	Voltage	v	volt
Thermal Resistivity	R	degree-Sec/calorie	Electrical Resistance	R	ohm
Thermal Capacity	C	calorie/degree	Capacitance	C	farad
Guiding Equations					
Thermal Conduction		$Q = \frac{1}{R}(T_i - T_j)$	Electric Conduction		$i = \frac{1}{R}(v_2 - v_1)$
Energy Storage (Thermal)		$\frac{dT}{dt} = \frac{1}{C}Q$	Energy Storage (Electrical)		$\frac{dv}{dt} = \frac{1}{C}i$

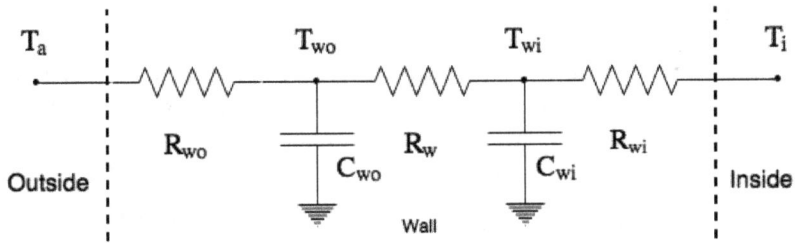

Fig. C.1   3R2C model of a single wall.

(2) Internal heat gain is constant.
(3) HVAC systems are ideal and work at constant efficiency such that they always meet the cooling and heating requirements for the designed temperature range at constant consumption.
(4) Radiative heat transfer formulation is linearized.

The RC model is built by synthesis of models of various elements of the building viz., walls, windows, floors, roofs and doors. A typical example [Belic *et al.* (2016)] of RC model is a 3R2C model of a single wall as shown in Figure C.1.

The 3R2C model captures the convection between outside air and the outer surface of the wall, conduction between the two layers of the wall and convection between the room air and the inner surface of the wall. With RC method, it is also possible to capture solar radiation and radiative heat from other sources like heaters, body heat, etc. for radiative heat exchanges that can be linearized as discussed in Section 4.4.1.

State-Space Representation

The state of a dynamic system is a set of physical quantities and the specification of these physical quantities completely determines how the system will evolve with time in the absence of any external excitation.

Newtonian calculus is used in order to characterize the behavior of a dynamic system. The behavior of dynamic systems is represented by systems of ordinary differential equations[8] and it is these differential equations,

---

[8]It may be noted that in the state-space approach [Friedland (2005)], all the differential equations in the mathematical model of a system are first-order equations: only the dynamic variables and their first derivatives with respect to time appear in the differential equations.

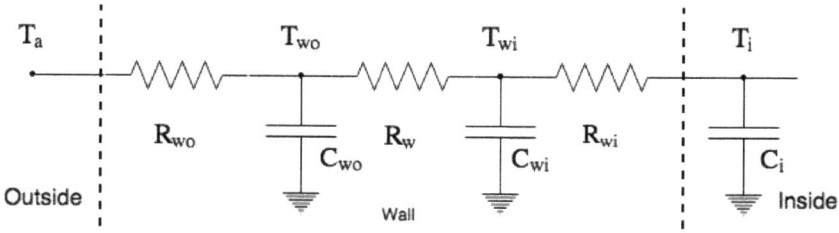

Fig. C.2   RC model of a system considering the effect of a single wall.

which constitute a mathematical model of a particular physical process. By solving the differential equations used in modeling a process, we can predict how the physical process will behave.

In order to develop the state-space representation of a simplified system of a room (considering the heat flow exchange through a single wall only) modeled as RC network as in Figure C.2, we need to formulate the differential equations. Note that a capacitance C_i (vis-a-vis the RC model of wall shown in Figure C.1) equivalent to the thermal capacity of the space inside the room is added for our simplified system and assumed that there is no heater or air-conditioner in the room.

Using nodal analysis,

$$C_{wo}\frac{dT_{wo}}{dt} - \frac{T_a - T_{wo}}{R_{wo}} + \frac{T_{wo} - T_{wi}}{R_w} = 0 \qquad (C.29)$$

$$C_{wi}\frac{dT_{wi}}{dt} - \frac{T_{wo} - T_{wi}}{R_w} + \frac{T_{wi} - T_i}{R_{wi}} = 0 \qquad (C.30)$$

$$C_i\frac{dT_i}{dt} - \frac{T_{wi} - T_i}{R_{wi}} = 0 \qquad (C.31)$$

Rearranging and using dot notation for derivatives,

$$d\dot{T}_{wo} + \frac{R_{wo} + R_w}{C_{wo}.R_{wo}.R_w}T_{wo} - \frac{1}{C_{wo}.R_w}T_{wi} + \frac{1}{C_{wo}.R_{wo}}T_a = 0 \quad (C.32)$$

$$\dot{T}_{wi} - \frac{1}{C_{wi}.R_w}T_{wo} + \frac{R_{wi}.R_w}{C_{wi}.R_{wi}.R_w}T_{wi}T_{wi} + \frac{1}{C_{wi}.R_{wi}}T_i = 0 \quad (C.33)$$

$$\dot{T}_i - \frac{1}{C_i.R_{wi}}T_{wi} + \frac{1}{C_i.R_{wi}}T_i = 0 \quad (C.34)$$

In matrix form, the system would be as shown in Equation (C.35).

$$\begin{bmatrix} \dot{T}_{wo} \\ \dot{T}_{wi} \\ \dot{T}_i \end{bmatrix} = \begin{bmatrix} \frac{R_{wo}+R_w}{C_{wo}.R_{wo}.R_w} & -\frac{1}{C_{wo}.R_w} & 0 \\ -\frac{1}{C_{wi}.R_w} & \frac{R_{wi}.R_w}{C_{wi}.R_{wi}.R_w}T_{wi} & \frac{1}{C_{wi}.R_{wi}} \\ 0 & -\frac{1}{C_i.R_{wi}} & \frac{1}{C_i.R_{wi}} \end{bmatrix} \begin{bmatrix} T_{wo} \\ T_{wi} \\ T_i \end{bmatrix} + \begin{bmatrix} \frac{1}{C_{wo}.R_{wo}} \\ 0 \\ 0 \end{bmatrix} [T_a]$$

$$(C.35)$$

The system described in Equation (C.35) can be presented in a typical state-space form of time-invariant system:

$$\mathbf{x}[k+1] = A.\mathbf{x}[k] + B.\mathbf{u}[k] \qquad (C.36)$$

where, $\mathbf{x}$ is a state vector, $\mathbf{u}$ is the input vector, A is state matrix and B is input matrix.

The output equation in state-space form would be

$$\mathbf{y}[k] = C.\mathbf{x}[k] \qquad (C.37)$$

where, $\mathbf{y}$ is the output vector and C is the output matrix. For example, if we just want to observe the effect of ambient temperature T_a (input) on the inside temperature T_i, the output equation will become

$$[T_i] = \begin{bmatrix} 0 & C & 1 \end{bmatrix} \begin{bmatrix} T_{wo} \\ T_{wi} \\ T_i \end{bmatrix} \qquad (C.38)$$

If we want to control the inside temperature T_i using a HVAC system for heating or cooling, as the requirement may be, we have to model the HVAC system and the vector $\mathbf{y}$ will have more element(s) owing to radiant heat from heater or convective heat transfer through the cooling coil of a window AC (heat transfer through chilled water pipes in case of chilled water system). It may be noted that from control point of view the system model described in Equation (C.35) is actually the model of the disturbance in the system and the job of the HVAC controller is to maintain the desired thermal comfort i.e., the inside temperature T_i.

Appendix D

Excerpts from IEC Standard 7730 for Calculation of PMV

The key information from the International Standards published by ASHRAE (American Society of Heating, Refrigerating and Air-Conditioning Engineers) and IEC (International Electrotechnical Commission) for calculating predicted mean vote (PMV) is presented here.

D.1 Calculating PMV

Calculate the PMV using Equations (D.1) to (D.4):

$$PMV = [0.303 exp(-0.036M) + 0.028] \times$$

$$\left\{ \begin{array}{r} (M - W) - 3.05 \times 10^{-3}[5733 - 6.99(M - W) - p_a] \\ -0.42[(M - W) - 58.15] - 1.7 \times 10^{-5}M(5867 - p_a) \\ -0.0014M(34 - t_a) - 3.96 \times 10^{-8}f_{cl}[(t_{cl} + 273)^4 - (\bar{t}_r + 273)^4] \\ -f_{cl}h_c(t_{cl} - t_a) \end{array} \right\} \tag{D.1}$$

$$t_{cl} = 35.7 - 0.028(M - W)$$
$$-I_{cl}\{3.96 \times 10^{-8}f_{cl}[(t_{cl} + 273)^4) - (\bar{t}_r + 273)^4)] + f_{cl}h_c(t_{cl} - t_a)\} \tag{D.2}$$

$$h_c = \begin{cases} 2.38 \, |t_{cl} - t_a|^{0.25} & \text{for } 2.38 \, |t_{cl} - t_a|^{0.25} > 12.1\sqrt{v_{air}} \\ 12.1\sqrt{v_{air}} & \text{for } 2.38 \, |t_{cl} - t_a|^{0.25} < 12.1\sqrt{v_{air}} \end{cases} \tag{D.3}$$

$$f_{cl} = \begin{cases} 1.00 + 1.290 I_{cl} & \text{for } I_{cl} \le 0.078 \, m^2\text{-}KW \\ 1.05 + 0.0645 I_{cl} & \text{for } I_{cl} > 0.078 \, m^2\text{-}KW \end{cases} \tag{D.4}$$

where

M is the metabolic rate, in watts per square meter (W/m^2);

W is the effective mechanical power, in watts per square meter (W/m^2);

I_{cl} is the clothing insulation, in square meters Kelvin per watt $(m^2 - K/W)$;

f_{cl} is the clothing surface area factor;

t_a is the air temperature, in degrees Celsius $(°C)$;

t_r is the mean radiant temperature, in degrees Celsius $(°C)$;

v_{air} is the relative air velocity, in meters per second (m/s);

p_a is the water vapour partial pressure, in pascals (Pa);

h_c is the convective heat transfer coefficient, in watts per square meter kelvin $[W/(m^2 - K)]$;

t_{cl} is the clothing surface temperature, in degrees Celsius $(°C)$.

PMV may be calculated for different combinations of metabolic rate, clothing insulation, air temperature, mean radiant temperature, air velocity and air humidity. The equations for t_{cl} and h_c may be solved by iteration.

The PMV index is usually derived for steady-state conditions but can be applied with good approximation during minor fluctuations of one or more of the variables, provided that time-weighted averages of the variables during the previous 1 h period are applied.

The index should be used only for values of PMV between -2 and $+2$, and when the six main parameters are within the following intervals:

M $46W/m^2$ to 232 W/m^2 (0.8 *met* to 4 *met*);

I_{cl} $0\ m^2\text{-}K/W$ to $0.310\ m^2\text{-}K/W$ (0 *clo* to 2 *clo*);

t_a $10°C$ to $30°C$;

$\bar{t}_r$ $10°C$ to $40°C$;

p_a $0\ Pa$ to $2700\ Pa$.
v_{air} $0\ m/s$ to $1m/s$;

D.2   Clothing Insulation Level

The clothing insulation level is shown in Table D.1.

Table D.1   Thermal insulation for typical combinations of garments (ISO Standard 7730:2005).

Work Clothing	I_{clo} clo	m^2-W	Daily wear clothing	I_{clo} clo	m^2-W
Underpants, boiler suit, socks, shoes	0.70	0.110	Panties, T-shirt, shorts, light socks, sandals	0.30	0.050
Underpants, shirt, boiler suit, socks, shoes	0.80	0.125	Underpants, shirt with short sleeves, light trousers, light socks, shoes	0.50	0.080
Underpants, shirt, trousers, smock, socks, shoes	0.90	0.140	Panties, petticoat, stockings, dress, shoes	0.70	0.105
Underwear with short sleeves and legs, shirt, trousers, jacket, socks, shoes	1.00	0.155	Underwear, shirt, trousers, socks, shoes	0.70	0.110
Underwear with long legs and sleeves, thermo-jacket, socks, shoes	1.20	0.185	Panties, shirt, trousers, jacket, socks, shoes	1.00	0.155
Underwear with short sleeves and legs, shirt, trousers, jacket, heavy quilted outer jacket and overalls, socks, shoes, cap, gloves	1.40	0.220	Panties, stockings, blouse, long skirt, jacket, shoes	1.10	0.170
Underwear with short sleeves and legs, shirt, trousers, jacket, heavy quilted outer jacket and overalls, socks, shoes	2.00	0.310	Underwear with long sleeves and legs, shirt, trousers, V-neck sweater, jacket, socks, shoes	1.30	0.200
Underwear with long sleeves and legs, thermo-jacket and trousers, Parka with heavy quitting, overalls with heavy quilting, socks, shoes, cap, gloves	2.55	0.395	Underwear with short sleeves and legs, shirt, trousers, vest, jacket, coat, socks, shoes	1.50	0.230

Appendix E

More on Grid Applications and Data Dissemination

The application of data dissemination techniques (presented in Chapter 2), in two more grid applications viz. i) State Estimation (SE) of power system and ii) Monitoring Coherent Group of Generators (MCGG) is discussed here.

E.1 Introduction to MCGG and SE

E.1.1 *Coherent Group of Generators and Islanding*

It has been observed that a group of generators close to the faults have a tendency to accelerate[9] much faster than those located far away. These group of generators have significant influence on the stability of the power system.

The power system can be divided into parts based on this group of generators, which can be considered as coherent group of generators (CGG) and can be represented as a single generator in the model for analysis. *Coherency is a phenomenon, which is observable and can be observed in a grid for a given disturbance, where certain generators tend to swing together following disturbances.* A CGG is a group, where generators oscillate with the same angular speed, and the same generator terminal/busbar voltages in the same direction — in a complex but constant ratio.

One of the strategies to deal with power system faults and avoiding cascading failure is controlled splitting of interconnected transmission network at proper points (by opening circuit breakers of selected transmission lines) to create islands of connected generator(s), which match with the

---

[9]Note that a generator accelerates, when its load decreases and a similar situation arises when circuit breakers open to isolate a fault disconnecting some loads in the process.

loads. The feasible splitting of power system in order to avoid blackout of the entire system is known as *islanding*. However, active islanding requires monitoring of coherent group of generators (MCGG). Monitoring is also required to alert the LDC (load dispatch centre) operator of possible islanding.

> *Monitoring Coherent Groups of Generators (MCGG)*: A group of generators is coherent if the difference between the centre of inertia of generators in the group from the global centre of inertia is within a given threshold. The difference beyond the thresholds is an alert to the system operator for possible islanding.

E.1.2   *SE of Power System*

State estimation (SE) of a power system is necessary to facilitate improved system reliability. The state variables in a power system are the voltage and current phasors (magnitude and phase angle) at all the buses (except the slack bus[10]) and it is necessary that the exact state of the power system is presented to the operator at regular intervals. Today's large and highly complex power systems require state estimator (SE) to support highly sophisticated techniques for monitoring (includes BAM and MCGG) and control of the system to maintain its reliability.

> *State Estimation (SE)*: Power system state estimation is a statistical technique to estimate the system state of a bus in the face of noisy and missing PMU data.

E.2   PSSE: Power System State Estimation

The state of a power system includes voltage magnitude and angle of buses. State estimation plays a vital role in various other grid applications. Though less time-critical compared to other grid applications, the operators rely heavily on the results of state estimation to determine the health of the grid. Before the advent of PMUs, the measurements (voltage and current) in a SCADA system were available from remote terminal

---

[10]A slack bus or swing bus is a bus connected to a large generator. This is used in electrical load flow analysis, where the imbalance between generation and demand of active and/or power is absorbed/supplied by this bus.

units. These measurements were not time synchronized which led to time skew errors. In addition to these errors, the measurements contain white noise. State estimation is a statistical method to obtain the state of the bus by minimizing the error in these measurements. If z and x are the measurement vector and the current state vector respectively then the model used in power system state estimation [Jones (2011)] is:

$$z = h(x) + e \qquad (E.1)$$

where h is the vector of nonlinear measurement functions between x and z, and e is the measurement white noise vector. The nonlinearity here is due to time skew errors in the measurements. In state estimation, i.e., estimating x from z, function h depends on admittance values in various buses. Performing state estimation involves solving non-convex optimization problems. However, typically, such models are iteratively linearized.

Due to the time synchronized measurements provided by the PMUs, the relation in (E.1) for an observable system is linear and can be solved in a single step. A system is observable if there are enough PMUs, such that the state estimator can estimate all the states of the grid. The method of estimating the states from the PMU measurements is known as Linear State Estimation (LSE). LSE reduces the computational complexity unlike nonlinear state estimation, which may take many iterations to converge towards a solution. The linear relation between the measurements and the state is [Navalkar (2012)]:

$$\mathbf{Z} = \mathbf{MV} + \epsilon \qquad (E.2)$$

where $\mathbf{Z}$ is the measurement vector, $\mathbf{M}$, the model matrix, which depends upon network connectivity and admittance values, $\mathbf{V}$ is the vector of voltage estimates at buses and ϵ indicates the white noise in the measurements.

As a transmission grid consists of many interconnected and geographically distributed regions (or areas), each region sending data to its LPDC can perform LSE of that region. For example, in the context of Indian electric grid, the PDC for Maharashtra grid can perform LSE using the phasor information from the data packets of all the substation PMUs in the state. However, globally optimal control action requires knowing the global state of the system. PSSE with CEUT involves performing LSE at SPDC using the phasor information extracted from the aggregated PMU data packets sent by LPDCs.

E.2.1 *PSSE with DEFT*

Instead of performing state estimation at SPDC using all the phasors data present in the data packets sent by downstream LPDCs, the distributed algorithm proposed in [Hossain *et al.* (2012)] use only the limited information sent by the LPDCs. The states estimated by the distributed algorithm are close to those estimated by LSE at SPDC. The state of each bus is obtained by performing the weighted average of the estimates from the state estimators running at each downstream LPDC. The weight associated with each estimate is also shared along with the estimate by the regional LPDCs to the central node to perform the weighted average of the states. Besides the reduction in communication traffic and latency at SPDC, another benefit of distributed PSSE is the privacy of the transmission grid [Hossain *et al.* (2012)]. Communication channels between LPDCs and SPDC are vulnerable to cyberattacks and hence, sending only state estimates of the network will hide the data of the producer and consumer of electric power from unauthorized entities.

We will now discuss the implementation of the distributed algorithm for state estimation concerning DEFT that comprises dividing the state estimation problem into sub-queries at LPDC, execution of those sub-queries at LPDCs, sending the sub-query results to SPDC, and the computations performed at SPDC. Consider two regions connected by a transmission line (also known as *tie line*) such that LPDC of each region performs LSE given by (E.2). The buses from both the regions that are connected by the tie line are known as *boundary buses* (refer to Figure 2.2 for the identification of boundary bus and tie lines). The state-specific data sent from LPDC of each region to SPDC comprises of different components. It includes non-boundary bus estimates computed using LSE, estimates of the state of the boundary buses of both regions, the current of the tie line connecting the boundary bus and the voltage at the boundary bus in the region, and the weights corresponding to estimates and measurements. Hence, for the network having two regions, the data packet from each LPDC contains the estimates of both the boundary buses that are connected by the tie line.

Unlike in [Hossain *et al.* (2012)] that obtains the global state estimate of each bus by performing a weighted average of the estimates, the SPDC in our simulation uses weighted least squares method to obtain the global estimates of the boundary buses. *The benefit of this method over weighted average is that it tries to minimize the error between the measurements and the estimations of these measurements corresponding to boundary buses.*

Further, as LSE at each region leads to two estimates per boundary bus that is available at SPDC, weighted least squares over these redundant estimates would give a unique estimate per boundary bus. This would improve the accuracy of the global state of the boundary buses. For the non-boundary buses, the estimates obtained from LSE at a region are unique and are retained as a global state of those buses.

As the amount of data transmitted from each LPDC to SPDC in PSSE with DEFT depends on the boundary buses of each region and the number of electric transmission lines connecting these regions, it varies for different grid topologies. To estimate the bandwidth used by the application with DEFT and CEUT in the two regions, r_1 and r_2, with both regions having n buses, b of which are the boundary buses, are connected by b transmission lines. Also, let us assume that both regions have l transmission lines in total. The meta info per PMU packet comprises of 16 bytes, and the size of each phasor is 8 bytes [IEEE Std C37.118.2]. Neglecting the meta info from PMU data, each LPDC in CEUT would send $8(n + l)$ bytes of data to SPDC, where 8 is the size of each phasor in bytes.

In DEFT, LSE of each region would estimate the states of the boundary buses of both regions. This would lead to $2b$ voltage estimates at each LPDC which also receives PMU data from PMUs located at b boundary buses. Thus, considering phasor data of 8 bytes, the measured and estimated values lead to a total of $24b$ bytes of data. The tie line currents measurements (b) at each LPDC lead to $8b$ bytes of data. Hence, the data sent to SPDC comprising of measurements (b voltages and b currents) and estimates ($2b$ voltages) is $32b$ bytes. Further, the weights associated with measurements and estimates lead to $16b$ bytes, where each weight value requires 4 bytes. Each regional LPDC also sends the voltage estimates of the non-boundary buses ($n - b$) to SPDC. Thus, the total data sent by each LPDC is

$$40b + 8n \quad \text{bytes.}$$

In comparison to CEUT, DEFT sends

$$\frac{40b + 8n}{8(n + l)} \quad \text{fraction of data to SPDC.}$$

For simplicity, let the two regions, r_1 and r_2 have the same configuration as Eastern region of the Indian electric grid given in Table 2.2. Accordingly, $n = 363$, $b = 5$ and $l = 1532$. In this case, *DEFT sends less than* 20% *data compared to CEUT* leading to a reduction in latency for PSSE.

E.2.2   *Comparison of PSSE with DEFT versus CEUT*

In this subsection, we first present the simulation environment for grid state estimation application on the same simulation setup used for BAM (mentioned in Section 2.6.2) and then give the observed processing latency for the application. We further give the estimated processing latency for the Indian electric grid. For the PSSE with DEFT, each LPDC (Eastern and Western regions) after performing LSE, sends the phasor estimates, measurements, and weights of the boundary buses to SPDC. The SPDC then performs weighted least squares from the measurements and estimates received from both LPDCs. The matrix computations involved in LSE and weighted least squares were performed with GNU Scientific Library.

E.2.2.1   *Latency for PSSE from PMU to SPDC*

The simulation results presented in Figure E.1 show that PSSE with DEFT had 34% reduction in average latency and had 81% reduction in bandwidth usage compared to CEUT. Out of 18.8 millisec of PMU to SPDC end-to-end latency for PSSE, $D_{deft}^{lpdc} + D_{deft}^{spdc}$ from Equations (2.9) and (2.10) contributed the highest, which was 16.4 millisec. The significant decrease in average latency of PSSE with DEFT is mainly attributed to the fact that the size of the model matrix at LPDCs was approximately half that of the one at SPDC in CEUT and very low processing latency due to the calculation of weighted least squares on a smaller size model matrix at SPDC. Similarly, in DEFT, two LPDCs sent just 2256 and 3068 bytes respectively compared to 15918 and 14782 bytes with CEUT, resulting in reduced bandwidth.

E.2.2.2   *Estimated Processing Latency for PSSE on the Indian Electric Grid*

We assume that the four regional PDCs (Northern, Eastern, Western and North Eastern) perform LSE and National level SPDC performs weighted least squares. This would require the state PDCs (including Maharashtra PDC) to parse and send the aggregated PMUs data to regional PDCs. Accordingly, the D_{ceut}^{lpdc} at Maharashtra PDC with PMU data of 2.5 KB is 20.8 millisec.

D^{parse} at Western region PDC (intermediate PDC) from Equation (2.9) is 12.3 millisec. As given in [Navalkar (2012)], the LSE performed on the above mentioned four regional grids of India together in 23 millisec. With the assumption that each regional PDC would take one-fourth of the processing required for LSE at National PDC, performing LSE at the regional PDC would take 5.75 millisec (D^{app}). Thus,

$$D_{deft}^{lpdc}=12.3 + 5.75 = 18.05 \text{ millisec.}$$

Since the size of the matrix at SPDC is much smaller compared to that required for LSE at LPDC, 5.75 millisec would be a strict upper bound on the execution time for weighted least squares. Considering weighted least squares processing latency ($D_{deft}^{spdc}=D^{app}$) as 5.75 millisec, the estimated processing latency for PSSE with DEFT in the Indian electric grid is

$$20.8 + 18.3 + 5.75 = 44.85 \text{ millisec.}$$

In comparison, the estimated processing latency for CEUT is 180.8 millisec. It includes the processing latency in Maharashtra PDC (D_{ceut}^{lpdc}), Western region PDC (D_{ceut}^{lpdc}), National PDC (D_{ceut}^{spdc}) of 20.8 millisec, 65 millisec and 95 millisec respectively with D^{app} of 23 millisec at National PDC to perform LSE.

PSSE with DEFT achieved a considerable reduction in latency in the simulation environment. However, reduction in processing latency with DEFT on Indian electric grid was quite significant. Further, in the simulation, DEFT required lesser bandwidth compared to CEUT.

E.3   MCGG: Monitoring Coherent Groups of Generators

Electric grids cover vast geographic areas. To study the stability of such systems, it is not practical to model the details of the entire grid. Rather, it is common practice to represent parts of the grid by equivalent models while preserving the general behavioural characteristics of the system. One step of creating these equivalent models is the identification of coherent groups of generators. When a *remote disturbance* occurs, coherent generators *swing* together and can, therefore, be represented by a single equivalent (generating) machine. Note that the notion of coherent group of generators is valid following a remote disturbance (fault) only and it is an implicit assumption for all the discussions under this section, unless mentioned otherwise.

When a system in the steady state is subjected to a disturbance, it undergoes electromechanical oscillations before settling down to a new steady state. These electromechanical oscillations which are commonly referred to as modes comprise characteristics such as an oscillatory frequency, damping, and wave spectrum. The mode shape describes the amplitude and phasing of the oscillation throughout the system. Accurate estimates of the electromechanical modes are required for safe and reliable operation of the grid. Each mode is characterized by the coherent groups of generators swinging against another coherent group of generators. A group of generators is coherent if all the generators in the group have similar response characteristics to variations in the operating states of a power system. There are different types of modes (local machine-systems, intra-plant mode, local mode, and inter-area mode) with different frequency of oscillations for the coherent groups of generators [Shubhanga and Anantholla (2006)]. The local machine-systems oscillations (0.7 to 2 Hz) include one or more synchronous machines at a power station swinging together against a larger power system or a load centre. The intra-plant mode oscillations (1.5 to 3 Hz) typically include more than two synchronous machines at a power plant swing against each other. The local mode oscillations (0.8 to 1.8 Hz) involve nearby power plants in which coherent groups of machines within an area swing against each other. The inter-area mode oscillations (0.1 to 0.5 Hz) normally involve many synchronous machines in one area of a power system swinging against machines in another area of the system. The modal analysis described in [Shubhanga and Anantholla (2006)] is used to determine the number of coherent groups in each mode, which coherent group swings against the other coherent group, which generators are part of each coherent group, etc. The modal analysis is a compute-intensive algorithm that involves solving eigenvalues and eigenvectors of large and sparse matrices. The implementation of this method in the simulation is beyond the scope of the monograph. However, it may be noted that MAT-LAB based package cited in [Shubhanga and Anantholla (2006)] is used for performing the modal analysis.

Of all the oscillation modes that were obtained from the modal analysis, some of the oscillations could be monitored at the local (LPDC) level. Here, we are monitoring coherent groups of generators at SPDC that belong to low-frequency inter-area oscillation mode. In real life, the configuration of the coherent groups of generators for an inter-area oscillation mode changes infrequently (typically, less than 5) in 24 hours.

Each coherent group of generators is associated with a centre of inertia (COI) that needs to be computed using the rotor angles of those generators. The phase angle of the generating substation can be used to derive the rotor angle of a generator. Here, we use a phase angle of a generating substation as the rotor angle of the generator. The COI for a coherent group of generators is the weighted average of rotor angles of the generators in the group. The weight associated with the angle of each generator is its inertia constant [Alsafih and Dunn (2010)]. If there are N groups in an inter-area swing mode with each group having $k_i, i \in [1, N]$ generators, the group i COI with k_i generators is given as in [Alsafih and Dunn (2010)]:

$$\delta_{COI_i} = \frac{\sum_{j=1}^{k_i} H_j * \delta_j}{G_i} \tag{E.3}$$

where H_j and δ_j are the inertia constant and rotor angle of generator j respectively. G_i is called the inertia constant of group i and is computed as $G_i = \sum_{j=1}^{k_i} H_j$.

When a small load trips (disturbance), there is an imbalance between cumulative generation and cumulative load and COI shifts. The system can remain in the equilibrium if the difference between the COI of a coherent group i, δ_{COI_i} from the global centre of inertia, δ_{COI} remains within the threshold values. The global COI is computed from group COI as follows:

$$\delta_{COI} = \frac{\sum_{i=1}^{n} G_i * \delta_{COI_i}}{W} \tag{E.4}$$

where δ_{COI_i} is COI of group i and $W = \sum_{i=1}^{n} G_i$.

MCGG implemented using CEUT comprises of computing group COIs and global COI at SPDC. SPDC then monitors the deviation of each group COI from global COI as

$$\epsilon_{l_i} \leq |\delta_{COI} - \delta_{COI_i}| \leq \epsilon_{u_i}, \forall i \in [1, N] \tag{E.5}$$

where $\epsilon_{l_i}, \epsilon_{u_i}$ are lower and upper bounds respectively for group i. The violation of condition (E.5) alerts the system operator about possible islanding. This requires extracting the phase angles of the generators involved in the coherent groups at SPDC from aggregated data received from LPDCs. The modal analysis discussed above can be performed to determine if the configuration of coherent groups has changed due to violation of (E.5).

E.3.1 *MCGG with DEFT*

Unlike CEUT, where group COIs and global COI are computed at SPDC, in DEFT, each LPDC computes the centre of inertia of the group of generators from which it receives data. The group COI is then sent to SPDC to calculate global COI. In the context of Indian electric grid [PGCIL (2012)], inter-area oscillation modes across the states are monitored at regional PDCs, and those across the regions are monitored at National PDC. The MCGG, specifically for Eastern and Western regions (that we have considered in our simulation) have modes with a coherent group of generators from one region swinging against the coherent group of generators in another region. If the computation of group COI for MCGG at one regional PDC does not depend on the PMU data of generators from other regions, it can be computed directly at the regional PDC and sent to SPDC. However, there can be low-frequency oscillation modes wherein some coherent groups that are dominated by generators of a region may have coherency with few generators of another region. In two of the three modes that are monitored at National PDC (which is the SPDC) in our simulation, generators at the boundary of the Eastern region were found to swing with generators of the Western region and vice-versa. Hence, for the distributed query execution of MCGG, to calculate group COI at LPDC (regional PDC in the simulation), PMU data of a few generators of one region (say Eastern) that formed a coherent group with the generators of the other region (say Western) was sent to LPDC of that region (Western).

Sub-query at LPDC to compute group i COI with k_i generators is given by (E.3). The group COIs from the LPDC is then sent to SPDC where the global COI is computed by (E.4). The SPDC then executes the query given in (E.5).

Consider all LPDCs monitoring the m modes, with each mode having n generators. In CEUT, each LPDC sends K_{mn} KB to SPDC in every 20 millisec, where K is the PMU data size in KB. In DEFT, $18 + 4m$ bytes are sent by each LPDC to SPDC, which includes, meta-info (16), statistics (2) and group COIs ($4m$). For $K = 0.1$ KB, $m = 3$ and $n = 20$, DEFT sends only 0.5% of the application data compared to CEUT, which leads to improved processing latency.

E.3.2   *Comparison of MCGG: DEFT versus CEUT*

For MCGG, we will not get into the detailed calculations of observed processing latency (obtained from the simulation environment) and the estimation of processing latency for the Indian electric grid. We will present only the results MCGG with DEFT vis-à-vis with CEUT in Figure E.1. This is because the same technique of DEFT is in reducing latency.

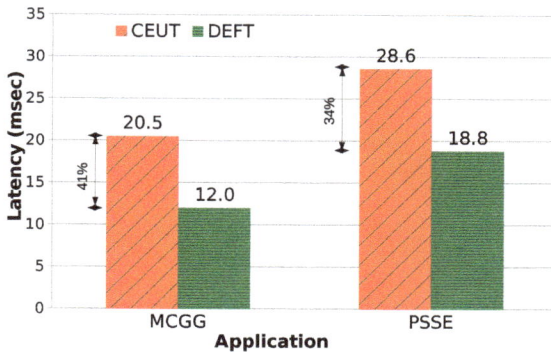

Fig. E.1 Latency reduction for MCGG and PSSE applications. [Khandeparkar *et al.* (2017)]

It can be observed from Figure E.1 that MCGG with DEFT has 63% of reduction in the processing latency over CEUT. It was found that MCGG with DEFT achieved significant reductions in latency both in the simulation environment and in a model for the Indian electric grid. Also, in the simulation, DEFT required lesser bandwidth compared to CEUT.

In this section, we present an extension to DEFT that takes cognizance priority of applications when multiple concurrent applications are executed on the same setup used in BAM (mentioned in Section 2.6.2). We also present extensions to this priority based DEFT that sends all the raw PMU data to SPDC.

Bibliography

Adamiak, M., Kasztenny, B., and Premerlani, W. (2005). Synchrophasors: Definition, measurement, and application, *Proceedings of the 59th Annual Georgia Tech Protective Relaying, Atlanta, GA*, pp. 27–29.

Aelenei, L., Pereira, R., Ferreira, A., Gonçalves, H., and Joyce, A. (2014). Building integrated photovoltaic system with integral thermal storage: A case study, *Energy Procedia* **58**, pp. 172–178, doi:10.1016/j.egypro.2014.10.425.

Agarwal, A. A., Munigala, V., and Ramamritham, K. (2016). Observability: A principled approach to provisioning sensors in buildings, in *Proceedings of the 3rd ACM International Conference on Systems for Energy-Efficient Built Environments (BuildSys '16)*. Association for Computing Machinery, New York, NY, USA, pp. 197–206. DOI:https://doi.org/10.1145/2993422.2993427.

Agarwal, Y., Balaji, B., Dutta, S., Gupta, R. K., and Weng, T. (2011). Duty-cycling buildings aggressively: The next frontier, in *HVAC Control, Proceedings of ACM Conference on Information Processing in Sensor Networks (IPSN)*.

Agarwal, Y., Balaji, B., Gupta, R., Lyles, J., Wei, M., and Weng, T. (2010). Occupancy-Driven Energy Management for Smart Building Automation, in *Proceedings of ACM Conference on Systems for Energy-Efficient Building Cities and Transportation (BuildSys)*.

Alsafih, H. and Dunn, R. (2010). Identification of critical areas for potential wide-area based control in complex power systems based on coherent clusters, *Universities Power Engineering Conference (UPEC)*.

Ardakanian, O., Bhattacharya, A., and Culler, D. (2016a). Non-intrusive techniques for establishing occupancy related energy savings in commercial buildings, in *Proceedings of the 3rd ACM International Conference on Systems for Energy-Efficient Built Environments*, BuildSys '16 (ACM, New York, NY, USA), ISBN 978-1-4503-4264-3, pp. 21–30, doi:10.1145/2993422. 2993574, `http://doi.acm.org/10.1145/2993422.2993574`.

Ardakanian, O., Bhattacharya, A., and Culler, D. (2016b). Non-intrusive techniques for establishing occupancy related energy savings in commercial buildings, in *Proceedings of the 3rd ACM International Conference on Systems for Energy-Efficient Built Environments* (ACM), pp. 21–30.

Armenia, A. and Chow, J. H. (2010). A flexible phasor data concentrator design leveraging existing software technologies, *IEEE Transactions on Smart Grid* **1**, 1, pp. 73–81.

ASHRAE (1992). *ASHRAE Standard 55-1992: Thermal Environmental Conditions for Human Occupancy* (American Society of Heating, Ventillation, Refrigerating and Air-Conditioning Engineers (ASHRAE), Atlanta, GA).

ASHRAE (2001). *Fundamentals of HVAC Systems* (ASHRAE Handbook, Atlanta, GA).

Babu, T. S., Ram, J. P., Dragičević, T., Miyatake, M., Blaabjerg, F., and Rajasekar, N. (2018). Particle swarm optimization based solar pv array reconfiguration of the maximum power extraction under partial shading conditions, *IEEE Transactions on Sustainable Energy* **9**, 1, pp. 74–85.

Bakken, D., Bose, A., Hauser, C., Whitehead, D., and Zweigle, G. (2011). Smart generation and transmission with coherent, real-time data, *Proceedings of the IEEE* **99**, 6, pp. 928 –951, doi:10.1109/JPROC.2011.2116110.

Balaji, B., Xu, J., Nwokafor, A., Gupta, R., and Agarwal, Y. (2013a). Sentinel: Occupancy based HVAC actuation using existing wifi infrastructure within commercial buildings, in *Proceedings of the 11th ACM Conference on Embedded Networked Sensor Systems*, SenSys '13 (ACM, New York, NY, USA), ISBN 978-1-4503-2027-6, pp. 17:1–17:14, doi:10.1145/2517351. 2517370, http://doi.acm.org/10.1145/2517351.2517370.

Balaji, B., Xu, J., Nwokafor, A., Gupta, R., and Agarwal, Y. (2013b). Sentinel: Occupancy based HVAC actuation using existing WiFi infrastructure within commercial buildings, in *ACM Conference on Embedded Networked Sensor Systems (SenSys)*.

Barker, S. K., Mishra, A. K., Irwin, D. E., Shenoy, P. J., and Albrecht, J. R. (2012). Smartcap: Flattening peak electricity demand in smart homes, in *Proceedings of the 2012 IEEE International Conference on Pervasive Computing and Communications*, pp. 67–75.

Basina, D. R., Kumar, S., Padhi, S., Sarkar, A., Mondal, A., and krithi, R. (2020a). Brownout based blackout avoidance strategies in smart grids, *IEEE Transactions on Sustainable Computing*, pp. 1–1, doi:10.1109/TSUSC.2020. 3014077.

Basina, D. R., Kumar, S., Padhi, S., Sarkar, A., Mondal, A., and krithi, R. (2020b). Brownout based blackout avoidance strategies in smart grids, *IEEE Transactions on Sustainable Computing*, pp. 1–1, doi:10.1109/ TSUSC.2020.3014077.

Belic, F., Hocenski, Z., and Sliskovic, D. (2016). Thermal modeling of buildings with RC method and parameter estimation, in *2016 International Conference on Smart Systems and Technologies (SST)*, pp. 19–25, doi: 10.1109/SST.2016.7765626.

Beltran, A., Erickson, V. L., and Cerpa, A. E. (2013). Thermosense: Occupancy thermal based sensing for HVAC control, in *BuildSys* (ACM), pp. 11:1–11:8.

Biljecki, F., Ledoux, H., and Stoter, J. (2016). An improved lod specification for 3d building models, *Computers, Environment and Urban Systems* **59**, pp. 25–37, doi:https://doi.org/10.1016/j.compenvurbsys.2016.04.005, https://www.sciencedirect.com/science/article/pii/S0198971516300436.

Bobba, R. B., Dagle, J., Heine, E., Khurana, H., Sanders, W. H., Sauer, P., and
Yardley, T. (2012). Enhancing grid measurements: Wide area measurement
systems, naspinet, and security, *IEEE Power and Energy Magazine* **10**, 1,
pp. 67–73, doi:10.1109/MPE.2011.943133.

Bose, R. K. (2010). Energy efficient cities: assessment tools and benchmarking
practices, Tech. rep., The World Bank, Washington, D.C.

Boyd, S., Kim., S.-J., Vandenberghe, L., and Hassibi, A. (2007). A tutorial on
geometric programming, *Optimization and Engineering* **8**, 67, https://
doi.org/10.1007/s11081-007-9001-7.

Brager, G. and de Dear, R. (2001). Climate, comfort and natural ventilation:
a new adaptive comfort standard for ashrae standard 55, *Indoor Environmental Quality Series, Proceedings, Moving Thermal Comfort Standards
into the 21st Century*.

Brager, G. S. and de Dear, R. J. (1998). Thermal adaptation in the built environment: a literature review, *Energy and Buildings* **27**, pp. 83–96.

Bueno, B., Norford, L., Pigeon, G., and Britter, R. (2012). A resistance-
capacitance network model for the analysis of the interactions between the
energy performance of buildings and the urban climate, *Building and Environment* **54**, pp. 116–125, doi:https://doi.org/10.1016/j.buildenv.2012.01.
023.

Busquets-Monge, S., Rocabert, J., Rodriguez, P., Alepuz, S., and Bordonau,
J. (2008). Multilevel diode-clamped converter for photovoltaic generators
with independent voltage control of each solar array, *IEEE Transactions
on Industrial Electronics* **55**, 7, pp. 2713–2723.

CEA-India (2013). Growth of electricity sector in India from 1947–2013, Tech.
rep., Central Electricity Authority.

Chenine, M. and Nordström, L. (2011). Modeling and simulation of wide-area
communication for centralized pmu-based applications, *IEEE Transactions
on Power Delivery* **26**, 3, pp. 1372–1380.

Chil Prakash, V., Prakash, A. K., Arote, U., Munigala, V., and Ramamritham,
K. (2015). Demo abstract: Demonstration of using sensor fusion for con-
structing a cost-effective smart-door, in *Proceedings of the 2015 ACM Sixth
International Conference on Future Energy Systems*, e-Energy '15 (ACM,
New York, NY, USA), ISBN 978-1-4503-3609-3, pp. 191–192, doi:10.1145/
2768510.2770938, http://doi.acm.org/10.1145/2768510.2770938.

Clear, A., Friday, A., Hazas, M., and Lord, C. (2014). Catch my drift?: Achieving
comfort more sustainably in conventionally heated buildings, in *Proceedings
of the 2014 Conference on Designing Interactive Systems*, DIS '14 (ACM,
New York, NY, USA), ISBN 978-1-4503-2902-6, pp. 1015–1024, doi:10.
1145/2598510.2598529, http://doi.acm.org/10.1145/2598510.2598529.

Colson, C. M. and Nehrir, M. H. (2009). A review of challenges to real-time power
management of microgrids, in *2009 IEEE Power Energy Society General
Meeting*, pp. 1–8, doi:10.1109/PES.2009.5275343.

Davies, M. G. (1983). Optimal designs for star circuits for radiant exchange in a
room, *Building and Environment* **18**, 3, pp. 135–150.

Davies, M. G. (1984). Lumping radiant and convective exchange in a room, *Building Services Engineering Research and Technology* **5**, 1, pp. 28–31.

de Dear, R. J. and Brager, G. S. (2002). Thermal comfort in naturally ventilated buildings: revisions to ASHRAE standard 55, *Energy and Buildings* **34**, pp. 549–561.

DoEE, A. G. (2017). *Heating, Ventilation and Air-Conditioning*, http://eex.gov.au/technologies/heating-ventilation-and-air-conditioning/.

Dong, H., Yan, X., Chao, F., and Li, Y. (2008). Predictive control model for radiant heating system based on neural network, in *2008 International Conference on Computer Science and Software Engineering*, pp. 45–48, doi: 10.1109/CSSE.2008.490.

Ekoe a Akata, M. A., Njomo, D., and Mempouo, B. (2015). The effect of building integrated photovoltaic system (bipvs) on indoor air temperatures and humidity (iath) in the tropical region of cameroon, *Future Cities and Environment* **1**.

Erickson, V., Carreira-Perpinan, M., and Cerpa, A. (2011). OBSERVE: Occupancy-based System for Efficient Reduction of HVAC Energy, in *IPSN*.

Eternal Sun (2018). Solar simulator, http://www.eternalsun.com/products/solar-simulator/, accessed on: August 22, 2018.

Fanger, P. (1973). Assessment of man's thermal comfort in practice, *British Journal of Industrial Medicine* **30**, pp. 313–324.

Fanger, P. O. (1970). *Thermal Comfort: Analysis and Applications in Environmental Engineering* (Danish Technical Press, Copenhagen, Denmark).

Farhangi, H. (2010). The path of the smart grid, *IEEE Power and Energy Magazine* **8**, 1, pp. 18–28.

Fitzgerald, A. E., Kingsley Jr., C., and Umans, S. D. (1985). *Electric Machinery (4th Ed.)* (McGraw-Hill Book Company, Singapore).

Fraisse, G., Viardot, C., Lafabrie, O., and Achard, G. (2002). Development of a simplified and accurate building model based on electrical analogy, *Energy and Buildings* **34**, 10, pp. 1017–1031.

Friedland, B. (2005). *Control Systems Design: An Introduction to State-Space Methods* (Dover Publications, Inc., Mineola, New York).

Gagge, A. P., Fobelets, A. P., and Berglund, L. G. (1986). A standard predictive index of human response to the thermal environment, *ASHRAE Transactions* **92**, 2B, pp. 709–731, http://oceanrep.geomar.de/42985/.

Gao, P. X. and Keshav, S. (2013a). SPOT: A smart personalized office thermal control system, in *The Fourth International Conference on Future Energy Systems*, e-Energy'13, pp. 237–246, doi:10.1145/2487166.2487193, http://doi.acm.org/10.1145/2487166.2487193.

Gao, P. X. and Keshav, S. (2013b). SPOT: A smart personalized office thermal control system, in *e-Energy* (ACM), pp. 237–246.

Ghai, S., Thanayankizil, L., Seetharam, D., and Chakraborty, D. (2012a). Occupancy detection in commercial buildings using opportunistic context sources, in *2012 IEEE International Conference on Pervasive Computing and Communications Workshops (PERCOM Workshops)*, pp. 463–466, doi: 10.1109/PerComW.2012.6197536.

Ghai, S. K., Thanayankizil, L. V., Seetharam, D. P., and Chakraborty, D. (2012b). Occupancy detection in commercial buildings using opportunistic context sources, in *2012 IEEE International Conference on Pervasive Computing and Communications Workshops (PERCOM Workshops)*, pp. 463–466.

Giordano, V. and Bossart, S. (2012). Assessing smart grid benefits and impacts: EU and U.S. initiatives joint report EC JRC – US DOE, Tech. rep., European Commission - JRC and US Department of Electronics, doi: 10.2790/63348(online).

Goyal, S., Liao, C., and Barooah, P. (2011). Identification of multi-zone building thermal interaction model from data, in *2011 50th IEEE Conference on Decision and Control and European Control Conference*, pp. 181–186, doi: 10.1109/CDC.2011.6161387.

Gupta, R. and Ramamritham, K. (2012). Query planning for continuous aggregation queries over a network of data aggregators, *IEEE Transactions on Knowledge and Data Engineering* **24**, 6, pp. 1065–1079.

Gupta, R., Ramamritham, K., and Mohania, M. (2010). Ratio threshold queries over distributed data sources, in *2010 IEEE 26th International Conference on Data Engineering (ICDE)*, pp. 581–584.

Hart, G. W. (1989). Residential energy monitoring and computerized surveillance via utility power flows, *IEEE Technology and Society Magazine* **8**, pp. 12–16.

Hart, G. W. (1992a). Nonintrusive appliance load monitoring, *Proceedings of the IEEE* **80**, 12, pp. 1870–1891, doi:10.1109/5.192069.

Hart, G. W. (1992b). Nonintrusive appliance load monitoring, *Proceedings of the IEEE* **80**, pp. 1870–1891.

Hazra, J., Das, K., Seetharam, D., and Singhee, A. (2011). Stream computing based synchrophasor application for power grids, in *Proceedings of the First International Workshop on High Performance Computing, Networking and Analytics for the Power Grid* (ACM), pp. 43–50.

Hossain, E., Han, Z., and Poor, V. (eds.) (2012). *Distributed State Estimation: A Learning Based Framework* (Cambridge University Press).

Hu, D. and Venkatasubramanian, V. (2007). New wide-area algorithms for detection and mitigation of angle instability using synchrophasors, in *2007 IEEE Power Engineering Society General Meeting*, pp. 1–8.

Huang, H., Chen, L., Mohammadzaheri, M., Hu, E., and Chen, M. (2013). Multi-zone temperature prediction in a commercial building using artificial neural network model, in *2013 10th IEEE International Conference on Control and Automation (ICCA)*, pp. 1896–1901, doi:10.1109/ICCA.2013.6565010.

Huizenga, C., Zhang, H., and Arens, E. (2001). A model of human physiology and comfort for assessing complex thermal environments, *Building and Environment* **36**, pp. 691–699.

IEEE Std 1159 (2019). IEEE recommended practice for monitoring electric power quality, *IEEE Std 1159-2019*.

IEEE Std 141 (1993). IEEE recommended practice for electrical power distribution for industrial plants, *IEEE Std 141-1993*, pp. 1–768, doi:10.1109/IEEESTD.1994.121642.

IEEE Std C37.118.2 (2011). IEEE standard for synchrophasor data transfer for power systeodbuss, Tech. rep., IEEE, doi:10.1109/IEEESTD.2011.6111222.

IEEE Std C37.244 (2013). IEEE guide for phasor data concentrator requirements for power system protection, control, and monitoring, *IEEE Std C37.244-2013*, pp. 1–65, doi:10.1109/IEEESTD.2013.6514039.

IFTTT (2017). Ifttt, `https://ifttt.com/`, accessed: 2017-09-13.

ISO (2005). *Ergonomics of the Thermal Environment — Analytical Determination and Interpretation of Thermal Comfort using Calculation of the PMV and PPD Indices and Local Thermal Comfort Criteria* (The International Organization for Standards, Case postale 56, CH-1211 Geneva 20).

ISO 16484 (2005). Iso 16484-6:2005, `https://www.iso.org/standard/37299.html`, accessed: 2017-09-13.

Iyengar, S., Kalra, S., Ghosh, A., Irwin, D. E., Shenoy, P. J., and Marlin, B. (2015). iprogram: Inferring smart schedules for dumb thermostats, in *BuildSys* (ACM), pp. 211–220.

Jain, A., Behl, M., and Mangharam, R. (2017). Data predictive control for building energy management, in *2017 American Control Conference (ACC)*, pp. 44–49, doi:10.23919/ACC.2017.7962928.

Jois, S., Ramamritham, K., and Agarwal, V. (2020a). Impact of facade based building integrated photovoltaics on indoor thermal conditions in Mumbai, in *2020 47th IEEE Photovoltaic Specialists Conference (PVSC)*, pp. 0636–0639, doi:10.1109/PVSC45281.2020.9300456.

Jois, S., Ramamritham, K., and Agarwal, V. (2020b). Optimization of facade based building integration photovoltaic module installation, in *2020 47th IEEE Photovoltaic Specialists Conference (PVSC)*, pp. 1502–1504, doi:10.1109/PVSC45281.2020.9300529.

Jones, K. D. (2011). *Three-Phase Linear State Estimation with Phasor Measurements*, Master's thesis, Virginia Tech University, Blacksburg, VA.

Karmakar, B. K. and Karmakar, G. (2021). A current supported pv array reconfiguration technique to mitigate partial shading, *IEEE Transactions on Sustainable Energy* **12**, 2, pp. 1449–1460, doi:10.1109/TSTE.2021.3049720.

Karmakar, B. K. and Pradhan, A. K. (2020). Detection and classification of faults in solar pv array using thevenin equivalent resistance, *IEEE Journal of Photovoltaics* **10**, 2, pp. 644–654.

Karmakar, G., Arote, U., Agarwal, A. A., and Ramamritham, K. (2018). Adaptive hybrid approaches to thermal modeling of building, in *Proceedings of the Ninth International Conference on Future Energy Systems, e-Energy 2018, Karlsruhe, Germany, June 12–15, 2018*, pp. 477–479, doi: 10.1145/3208903.3212068, `https://doi.org/10.1145/3208903.3212068`.

Karmakar, G., Kabra, A., and Ramamritham, K. (2015a). Maintaining thermal comfort in buildings: feasibility, algorithms, implementation, evaluation, *Real-Time Systems* **51**, 5, pp. 485–525, doi:10.1007/s11241-015-9231-2, `http://dx.doi.org/10.1007/s11241-015-9231-2`.

Karmakar, G., Kabra, A., and Ramamritham, K. (2015b). Maintaining thermal comfort in buildings: feasibility, algorithms, implementation, evaluation, *Real-Time Systems* **51**, 5, pp. 485–525.

Khandeparkar, K., Ramamritham, K., and Gupta, R. (2017). Qos-driven data processing algorithms for smart electric grids, *ACM Transactions on Cyber-physiccal Systems (TCPS)* **1**, 3, pp. 14:1–14:24, doi:10.1145/3047410, `http://doi.acm.org/10.1145/3047410`.

Kleiminger, W., Beckel, C., Staake, T., and Santini, S. (2013a). Occupancy detection from electricity consumption data, in *Proceedings of the 5th ACM Workshop on Embedded Systems For Energy-Efficient Buildings*, BuildSys'13 (ACM, New York, NY, USA), ISBN 978-1-4503-2431-1, pp. 10:1–10:8, doi:10.1145/2528282.2528295, `http://doi.acm.org/10.1145/2528282.2528295`.

Kleiminger, W., Beckel, C., Staake, T., and Santini, S. (2013b). Occupancy detection from electricity consumption data, in *BuildSys*.

Kleiminger, W., Santini, S., and Mattern, F. (2014). Smart heating control with occupancy prediction: How much can one save? in *Proceedings of the 2014 ACM International Joint Conference on Pervasive and Ubiquitous Computing: Adjunct Publication*, UbiComp '14 Adjunct (ACM, New York, NY, USA), ISBN 978-1-4503-3047-3, pp. 947–954, doi:10.1145/2638728.2641555, `http://doi.acm.org/10.1145/2638728.2641555`.

Kolbe, T. (2009). *Representing and Exchanging 3D City Models with CityGML* (Springer, Berlin, Heidelberg), `https://doi.org/10.1007/978-3-540-87395-2_2`.

Lam, A. H., Yuan, Y., and Wang, D. (2014). An occupant-participatory approach for thermal comfort enhancement and energy conservation in buildings, in *Proceedings of ACM Conference on Future Energy Systems (e-Energy)*, pp. 133–143.

Lariviere, I. and Lafrance, G. (1999). Modelling the electricity consumption of cities: effect of urban density, *Energy Economics* **21**, 1, pp. 53–66, `https://ideas.repec.org/a/eee/eneeco/v21y1999i1p53-66.html`.

Levenberg, K. (1944). A method for the solution of certain non-linear problems in least squares, *Quarterly of Applied Mathematics*.

Lian, K., Jhang, J., and Tian, I. (2014). A maximum power point tracking method based on perturb-and-observe combined with particle swarm optimization, *IEEE Journal of Photovoltaics* **4**, 2, pp. 626–633.

Liu, C. L. and Layland, J. W. (1973). Scheduling algorithms for multiprogramming in a hard-real-time environment, *J. ACM* **20**, 1, pp. 46–61, doi: http://doi.acm.org/10.1145/321738.321743.

Liu, J. W. S. (2000). *Real-Time Systems* (Pearson Education).

Liu, J. W. S. and Wei-Kuan, S. (1995). Algorithms for scheduling imprecise computations with timing constraints to minimize maximum error, *IEEE Transactions on Computers* **44**, 3, pp. 466–471.

Lopez, C. S. P. and Sangiorgi, M. (2014). Comparison assessment of bipv facade semi-transparent modules: Further insights on human comfort conditions, *Energy Procedia* **48**, pp. 1419–1428, doi:10.1016/j.egypro.2014.02.160, `http://www.sciencedirect.com/science/article/pii/S1876610214004226`.

Lu, J., Sookoor, T., Srinivasan, V., Gao, G., Holben, B., Stankovic, J., Field, E., and Whitehouse, K. (2010). The smart thermostat: Using occupancy sensors to save energy in homes, in *SenSys*.

Maharashtra State Electricity Distribution Company, India (2014). HT bill format, http://www.mahadiscom.in/, [Accessed 15-August-2014].

Melfi, R., Rosenblum, B., Nordman, B., and Christensen, K. (2011). Measuring building occupancy using existing network infrastructure, in *Proceedings of the 2011 International Green Computing Conference and Workshops*, IGCC '11 (IEEE Computer Society, Washington, DC, USA), ISBN 978-1-4577-1222-7, pp. 1–8, doi:10.1109/IGCC.2011.6008560, http://dx.doi.org/10.1109/IGCC.2011.6008560.

Michael Kerrisk, J. L., and Zijlstra, P. (2014). Linux programmer's manual, http://man7.org/linux/man-pages/man7/sched.7.html.

Milenkovic, M. and Amft, O. (2013). An opportunistic activity-sensing approach to save energy in office buildings, in *Proceedings of the Fourth International Conference on Future Energy Systems*, e-Energy '13 (ACM, New York, NY, USA), ISBN 978-1-4503-2052-8, pp. 247–258, doi:10.1145/2487166.2487194, http://doi.acm.org/10.1145/2487166.2487194.

Modbus (2012). Modbus application protocol specification v1.1b3, Tech. rep., Modbus Organization, Modbus Organization, Inc. PO Box 628 Hopkinton, MA 01748.

Modbus Protocol (1996). Modicon modbus protocol reference guide — pi–mbus–300 rev. j, Tech. rep., Modicon Inc., One High Street North Andover, Massachusetts 01845.

Modbus TCP/IP (2006). Modbus messaging on tcp/ip implementation guide v1.0b, Tech. rep., Modbus Organization, Modbus Organization, Inc. PO Box 628 Hopkinton, MA 01748.

Nalmpantis, C. and Vrakas, D. (2019). Machine learning approaches for non-intrusive load monitoring: From qualitative to quantitative comparison, *Artif. Intell. Rev.* **52**, pp. 217–243.

Navalkar, P. V. (2012). *Phasor Measurement Unit Based Linear State Estimate Or-Diagnostics And Application to Secure Remote Backup Protection of Transmission Lines*, Ph.D. thesis, Department of Electrical Engineering, IIT Bombay, India.

Nelkon, M. and Parker, P. (1977). *Advanced Level Physics (4th Ed.)* (Heinemann Educational Books Limited, London).

Newsham, G. and Birt, B. (2010). Building-level occupancy data to improve ARIMA-based electricity use forecasts, in *BuildSys*.

Nichols, B., Buttlar, D., and Farrell, J. P. (1996). *Pthreads Programming* (O'Reilly & Associates, Inc., Sebastopol, CA, USA), ISBN 1-56592-115-1.

Olston, C., Jiang, J., and Widom, J. (2003). Adaptive filters for continuous queries over distributed data streams, in *Proceedings of the 2003 ACM SIGMOD International Conference on Management of Data* (ACM), pp. 563–574.

Parsons, K. (2003). *Human Thermal Environments* (CRC Press, London), doi: https://doi.org/10.1201/9780203302620.

PGCIL (2012). Unified real time dynamic state measurement (urtdsm).

Phadke, A. (1993). Synchronized phasor measurements in power systems, *IEEE Computer Applications in Power* **6**, 2, pp. 10–15.

Phadke, A. G. and Thorp, J. S. (2010). Communication needs for wide area measurement applications, in *2010 5th International Conference on Critical Infrastructure (CRIS)*, pp. 1–7, doi:10.1109/CRIS.2010.5617484.

Pisharoty, D., Yang, R., Newman, M. W., and Whitehouse, K. (2015). Thermocoach: Reducing home energy consumption with personalized thermostat recommendations, in *BuildSys* (ACM), pp. 201–210.

Power, T. (2014). Tata power tariff, `https://cp.tatapower.com/`, [Accessed 15-August-2014].

Prakash, A., Prakash, V., Doshi, B., Arote, U., Sahu, P., and Ramamritham, K. (2015). Fusing sensors for occupancy sensing in smart buildings, in *Proceedings of the International Conference on Distributed Computing and Intelligent Technology (ICDIT)*.

Ramanujam, A., Parihar, M., Swain, S., and Ramamritham, K. (2020). Design and development of brownout control strategy using end-point load control, in *Proceedings of the Eleventh ACM International Conference on Future Energy Systems*, e-Energy '20 (Association for Computing Machinery, New York, NY, USA), ISBN 9781450380096, pp. 293–298, doi:10.1145/3396851. 3397738, `https://doi.org/10.1145/3396851.3397738`.

Rizzo, S. A. and Scelba, G. (2015). ANN based MPPT method for rapidly variable shading conditions, *Applied Energy* **145**, pp. 124–132.

Rogers, G. F. C. and Mayhew, Y. R. (1982). *Engineering Thermodynamics: Work and Heat Transfer (3rd Ed.)* (ELBS and Longman Group Limited, London).

Saha, M. and Srivastava, B. N. (1967). *A Text Book of Heat (12th Ed.)* (Science Book Agency, Calcutta).

Salameh, Z. M. and Dagher, F. (1990). The effect of electrical array reconfiguration on the performance of a pv-powered volumetric water pump, *IEEE Transactions on Energy Conversion* **5**, 4, pp. 653–658.

Schweiker, M., Huebner, G. M., Kingma, B. R. M., Kramer, R., and Pallubinsky, H. (2018). Drivers of diversity in human thermal perception — a review for holistic comfort models, *Temperature* **5**, 4, pp. 308–342, doi: 10.1080/23328940.2018.1534490, `https://doi.org/10.1080/23328940.2018.1534490`, pMID: 30574525.

Seyedmahmoudian, M., Soon, T. K., Horan, B., Ghandhari, A., Mekhilef, S., and Stojcevski, A. (2019). New armo-based mppt technique to minimize tracking time and fluctuation at output of pv systems under rapidly changing shading conditions, *IEEE Transactions on Industrial Informatics*, pp. 1–1.

Shan, K., Wang, S., Gao, D., and Xiao, F. (2016). Development and validation of an effective and robust chiller sequence control strategy using data-driven models, *Automation in Construction* **65**, pp. 78–85, doi:https://doi.org/10.1016/j.autcon.2016.01.005, `http://www.sciencedirect.com/science/article/pii/S0926580516300097`.

Sharma, P. and Agarwal, V. (2014). Maximum power extraction from a partially shaded pv array using shunt-series compensation, *IEEE Journal of Photovoltaics* **4**, 4, pp. 1128–1137.

Shubhanga, K. and Anantholla, Y. (2006). Manual for a multi-machine small-signal stability programme, *Dept. of Electrical Engg. NITK, Surathkal.*

Singh, R. and Banerjee, R. (2015). Estimation of rooftop solar photovoltaic potential of a city, *Solar Energy*, doi:10.1016/j.solener.2015.03.016.

Skruch, P. (2015). A thermal model of the building for the design of temperature control algorithms, *Automatyka/Automatics* **18**, 1.

Subramanian, A., Garcia, M., Dominguez-Garcia, A., Callaway, D., Poolla, K., and Varaiya, P. (2012). Real-time scheduling of deferrable electric loads, in *American Control Conference (ACC), 2012* (IEEE), pp. 3643–3650.

Sun, Y., Wang, S., and Xiao, F. (2013). In situ performance comparison and evaluation of three chiller sequencing control strategies in a super high-rise building, *Energy and Buildings* **61**, pp. 333–343, doi:https://doi.org/10.1016/j.enbuild.2013.02.043, http://www.sciencedirect.com/science/article/pii/S0378778813001357.

Sundareswaran, K., Sankar, P., Nayak, P., Simon, S. P., and Palani, S. (2015). Enhanced energy output from a PV system under partial shaded conditions through artificial bee colony, *IEEE Transactions on Sustainable Energy* **6**, 1, pp. 198–209.

Sundareswaran, K., Sankar, P., Nayak, P. S. R., Simon, S. P., and Palani, S. (2015). Enhanced energy output from a pv system under partial shaded conditions through artificial bee colony, *IEEE Transactions on Sustainable Energy* **6**, 1, pp. 198–209.

Tapia, E., Intille, S., and Larson, K. (2004). Activity recognition in the home setting using simple and ubiquitous sensors, in *Pervasive*.

Ting, K., Yu, R., and Srivastava, M. (2013). Occupancy inferencing from non-intrusive data sources, in *BuildSys*.

Trivedi, A., Gummeson, J., Irwin, D., Ganesan, D., and Shenoy, P. (2017). ischedule: Campus-scale hvac scheduling via mobile wifi monitoring, in *Proceedings of the Eighth International Conference on Future Energy Systems*, e-Energy '17 (ACM, New York, NY, USA), ISBN 978-1-4503-5036-5, pp. 132–142, doi:10.1145/3077839.3077846, http://doi.acm.org/10.1145/3077839.3077846.

Vanfretti, L. and Chow, J. H. (2011). Synchrophasor data applications for wide-area systems, in *17th Power Systems Computation Conference (PSCC), Stockholm, Sweden, 2011.*

Vedova, M. L. D., Palma, E. D., and Facchinetti, T. (2011). Electric loads as real-time tasks: An application of real-time physical systems, in *Proceedings of the 7th International Wireless Communications and Mobile Computing Conference, IWCMC 2011*, pp. 1117–1123.

Vedova, M. L. D., Ruggieri, M., and Facchinetti, T. (2010). On real-time physical systems, in *18th International Conference on Real-Time and Network Systems RTNS, 2010*, pp. 41–49.

Velasco-Quesada, G., Guinjoan-Gispert, F., Piqué-López, R., Román-Lumbreras, M., and Conesa-Roca, A. (2009). Electrical PV array reconfiguration strategy for energy extraction improvement in grid-connected PV systems, *IEEE Transactions on Industrial Electronics* **56**, 11, pp. 4319–4331.

Wang, Y., Tian, W., Ren, J., Zhu, L., and Wang, Q. (2006). Influence of a building's integrated-photovoltaics on heating and cooling loads, *Applied Energy* **83**, 9, pp. 989–1003, doi:https://doi.org/10.1016/j.apenergy.2005.10.002, `https://www.sciencedirect.com/science/article/pii/S0306261905001339`.

William, D. and Stevenson, J. (1982). *Elements of Power System Analysis (4th Ed.)* (McGraw-Hill Book Company, Singapore).

Yang, T. and Athienitis, A. K. (2014). A study of design options for a building integrated photovoltaic/thermal (bipv/t) system with glazed air collector and multiple inlets, *Solar Energy* **104**, pp. 82–92, doi:https://doi.org/10.1016/j.solener.2014.01.049, `https://www.sciencedirect.com/science/article/pii/S0038092X14000899`, solar heating and cooling.

Young, H. D. and Freedman, R. A. (2008). *University Physics (12th Ed.)* (Pearson Addison Wesley, San Francisco, CA).

Zheng, Z., Chen, Q., Fan, C., Guan, N., Vishwanath, A., Wang, D., and Liu, F. (2018). Data driven chiller sequencing for reducing hvac electricity consumption in commercial buildings, in *Proceedings of the Ninth International Conference on Future Energy Systems*, e-Energy '18 (ACM, New York, NY, USA), ISBN 978-1-4503-5767-8, pp. 236–248, doi:10.1145/3208903.3208913, `http://doi.acm.org/10.1145/3208903.3208913`.

Zhu, K., Song, J., Chenine, M., and Nordstrom, L. (2010). Analysis of phasor data latency in wide area monitoring and control systems, in *2010 IEEE International Conference on Communications Workshops (ICC)*, pp. 1–5.

Index

About the Authors

Prof. Krithi Ramamritham has spent almost equal lengths of time at the University of Massachusetts, Amherst, and at IIT Bombay as a Chair Professor in the Department of Computer Science and Engineering. He is now a Distinguished Professor with Sai University, Chennai, India.

During 2006–2009, he served as Dean (R&D) at IIT Bombay. He also headed IIT Bombay's Center for Urban Science and Engineering (CUSE).

He holds a PhD degree in Computer Science from the University of Utah. He got his B.Tech (Electrical Engineering) and M.Tech (Computer Science) degrees from IIT Madras.

He is a recipient of IIT Bombay's prestigious S C Sahasrabuddhe Lifetime Achievement Award. He received the Distinguished Alumnus Award and the Robert Bosch Centre's Distinguished Fellowship from IIT Madras.

He is a Fellow of the IEEE, ACM, Indian Academy of Sciences, National Academy of Sciences, India, and the Indian National Academy of Engineering. He was honored with Outstanding Technical Contributions and Leadership Award from the IEEE Technical Committee for Real-Time Systems and the Outstanding Service Award from IEEE's CEDA.

His current research involves applying computational approaches to energy management, based on his S M A R T principle: *Sense Meaningfully, Analyze and Respond Timely*. As with his research since his PhD, his current work is driven by practical considerations and exploits and extends the state of the art in database systems, real-time computing, sensor networks, embedded systems, mobile computing and smart grids. He has guided over 40 PhD students. As per Google scholar, his publications have garnered over 23K citations with *h-index exceeding 80* (https://www.cse.iitb.ac.in/ krithi/publications.html).

Dr Gopinath Karmakar is a Scientific Officer-H at Bhabha Atomic Research Centre (BARC), India and also a faculty in BARC Training School. He has more than 30 years of experience in the field of Instrumentation and Control (I&C) for Safety-Critical Systems in nuclear power plants (NPP) and nuclear research reactors, which includes development of Hard Real-Time Systems, Operating System for safety-critical applications, Software Engineering for Class IA and IB systems, Programmable Controllers and System Engineering.

His present area of research is targeted towards Smart Building/Home and working in this field over the past few years. Dr Karmakar has served as a member of the program committee in various national and international conferences including ACM e-Energy.

Prashant Shenoy is a Distinguished Professor of Computer Science at the University of Massachusetts Amherst USA. His research interests are in distributed systems, networking, and computational sustainability. His work has won numerous best paper awards and a test of time award from ACM Sigmetrics. He serves as the founding chair of ACM Special Interest Group on Energy (SIGEnergy) and is on the editorial board of ACM Trans on Internet of Things. He has chaired energy-related conferences such as ACM e-Energy, ACM Buildsys and ACM/IEEE IoTDI. He is a Fellow of the ACM, the IEEE and the AAAS.